Study Guide for
Reteaching and Practice

by Cleo M. Meek

Algebra

Structure and Method   Book 1

Brown
Dolciani
Sorgenfrey
Cole

McDougal Littell
A HOUGHTON MIFFLIN COMPANY
EVANSTON, ILLINOIS
BOSTON ◆ DALLAS ◆ PHOENIX

SO-BZU-276

2004 Impression

Printed in the U.S.A.
ISBN: 0-395-47053-6

24 25-CSM-07 06 05

Contents

Symbols   vii

Table of Measures   viii

1 Introduction to Algebra

1-1	Variables	1-2
1-2	Grouping Symbols	3-4
1-3	Equations	5-6
1-4	Translating Words into Symbols	7-8
1-5	Translating Sentences into Equations	9-10
1-6	Translating Problems into Equations	11-12
1-7	A Problem Solving Plan	13-14
1-8	Number Lines	15-16
1-9	Opposites and Absolute Values	17-18

2 Working with Real Numbers

2-1	Basic Assumptions	19-20
2-2	Addition on a Number Line	21-22
2-3	Rules for Addition	23-24
2-4	Subtracting Real Numbers	25-26
2-5	The Distributive Property	27-28
2-6	Rules for Multiplication	29-30
2-7	Problem Solving: Consecutive Integers	31-32
2-8	The Reciprocal of a Real Number	33-34
2-9	Dividing Real Numbers	35-36

3 Solving Equations and Problems

3-1	Transforming Equations: Addition and Subtraction	37-38
3-2	Transforming Equations: Multiplication and Division	39-40
3-3	Using Several Transformations	41-42
3-4	Using Equations to Solve Problems	43-44
3-5	Equations with the Variable on Both Sides	45-46
3-6	Problem Solving: Using Charts	47-48
3-7	Cost, Income, and Value Problems	49-50
3-8	Proof in Algebra	51-52

4 Polynomials

4-1	Exponents	53-54
4-2	Adding and Subtracting Polynomials	55-56
4-3	Multiplying Monomials	57-58
4-4	Powers of Monomials	59-60
4-5	Multiplying Polynomials by Monomials	61-62
4-6	Multiplying Polynomials	63-64
4-7	Transforming Formulas	65-66
4-8	Rate-Time-Distance Problems	67-68
4-9	Area Problems	69-70
4-10	Problems without Solutions	71-72

5 Factoring Polynomials

5-1	Factoring Integers	73–74
5-2	Dividing Monomials	75–76
5-3	Monomial Factors of Polynomials	77–78
5-4	Multiplying Binomials Mentally	79–80
5-5	Differences of Two Squares	81–82
5-6	Squares of Binomials	83–84
5-7	Factoring Pattern for $x^2 + bx + c$, c positive	85–86
5-8	Factoring Pattern for $x^2 + bx + c$, c negative	87–88
5-9	Factoring Pattern for $ax^2 + bx + c$	89–90
5-10	Factoring by Grouping	91–92
5-11	Using Several Methods of Factoring	93–94
5-12	Solving Equations by Factoring	95–96
5-13	Using Factoring to Solve Problems	97–98

6 Fractions

6-1	Simplifying Fractions	99–100
6-2	Multiplying Fractions	101–102
6-3	Dividing Fractions	103–104
6-4	Least Common Denominators	105–106
6-5	Adding and Subtracting Fractions	107–108
6-6	Mixed Expression	109–110
6-7	Polynomial Long Division	111–112

7 Applying Fractions

7-1	Ratios	113–114
7-2	Proportions	115–116
7-3	Equations with Fractional Coefficients	117–118
7-4	Fractional Equations	119–120
7-5	Percents	121–122
7-6	Percent Problems	123–124
7-7	Mixture Problems	125–126
7-8	Work Problems	127–128
7-9	Negative Exponents	129–130
7-10	Scientific Notation	131–132

8 Introduction to Functions

8-1	Equations in Two Variables	133–134
8-2	Points, Lines, and Their Graphs	135–136
8-3	Slope of a Line	137–138
8-4	The Slope-Intercept Form of a Linear Equation	139–140
8-5	Determining an Equation of a Line	141–142
8-6	Functions Defined by Tables and Graphs	143–144
8-7	Functions Defined by Equations	145–146
8-8	Linear and Quadratic Functions	147–148
8-9	Direct Variation	149–150
8-10	Inverse Variation	151–152

9   Systems of Linear Equations

9-1	The Graphing Method	153–154
9-2	The Substitution Method	155–156
9-3	Solving Problems with Two Variables	157–158
9-4	The Addition-or-Subtraction Method	159–160
9-5	Multiplication with the Addition-or-Subtraction Method	161–162
9-6	Wind and Water Current Problems	163–164
9-7	Puzzle Problems	165–166

10   Inequalities

10-1	Order of Real Numbers	167–168
10-2	Solving Inequalities	169–170
10-3	Solving Problems Involving Inequalities	171–172
10-4	Solving Combined Inequalities	173–174
10-5	Absolute Value in Open Sentences	175–176
10-6	Absolute Values of Products in Open Sentences	177–178
10-7	Graphing Linear Inequalities	179–180
10-8	Systems of Linear Inequalities	181–182

11   Rational and Irrational Numbers

11-1	Properties of Rational Numbers	183–184
11-2	Decimal Forms of Rational Numbers	185–186
11-3	Rational Square Roots	187–188
11-4	Irrational Square Roots	189–190
11-5	Square Roots of Variable Expressions	191–192
11-6	The Pythagorean Theorem	193–194
11-7	Multiplying, Dividing, and Simplifying Radicals	195–196
11-8	Adding and Subtracting Radicals	197–198
11-9	Multiplication of Binomials Containing Radicals	199–200
11-10	Simple Radical Equations	201–202

12   Quadratic Functions

12-1	Quadratic Equations with Perfect Squares	203–204
12-2	Completing the Square	205–206
12-3	The Quadratic Formula	207–208
12-4	Graphs of Quadratic Equations: The Discriminant	209–210
12-5	Methods of Solution	211–212
12-6	Solving Problems Involving Quadratic Equations	213–214
12-7	Direct and Inverse Variation Involving Squares	215–216
12-8	Joint and Combined Variation	217–218

Looking Ahead

Sample Spaces and Events   219–220
Probability   221–222
Frequency Distributions   223–224
Presenting Statistical Data   225–226
Points, Lines, and Angles   227–228
Pairs of Angles   229–230
Triangles   231–232
Similar Triangles   233–234
Trigonometric Ratios   235–236
Values of Trigonometric Ratios   237–238
Problem Solving Using Trigonometry   239–240

Glossary of Properties

                                                  241–242

Symbols

		Page
·	× (times)	1
=	equals, is equal to	1
≠	is not equal to	1
( )	parentheses—a grouping symbol	1
[ ]	brackets—a grouping symbol	3
π	pi, a number approximately equal to $\frac{22}{7}$	
∈	is a member of, belongs to	5
∴	therefore	
$\overset{?}{=}$	is this statement true?	13
. . .	and so on	15
−	negative	15
+	positive	15
<	is less than	15
>	is greater than	15
^-a	opposite or additive inverse of a	17
$\|a\|$	absolute value of a	17
$\frac{1}{b}$	reciprocal or multiplicative inverse of b	33
∅	empty set, null set,	46
$a{:}b$	ratio of a to b	113

		Page
(a, b)	ordered pair whose first component is a and second component is b	133
$f(x)$	f of x, the value of f at x	145
≥	is greater than or equal to	167
≤	is less than or equal to	167
∩	the intersection of	
∪	the union of	
$\sqrt{}$	principal square root	187
≈	is approximately equal to	190
$P(A)$	probability of event A	221
$\overleftrightarrow{AB}$	line AB	227
$\overline{AB}$	segment AB	227
AB	the length of $\overline{AB}$	227
$\overrightarrow{AB}$	ray AB	227
∠	angle	227
°	degree(s)	227
Δ	triangle	231
~	is similar to	233
cos A	cosine of A	235
sin A	sine of A	235
tan A	tangent of A	235

Table of Measures

Metric Units

Length
$$10 \text{ millimeters (mm)} = 1 \text{ centimeter (cm)}$$
$$\left.\begin{array}{r}100 \text{ centimeters}\\1000 \text{ millimeters}\end{array}\right\} = 1 \text{ meter (m)}$$
$$1000 \text{ meters} = 1 \text{ kilometer (km)}$$

Area
$$100 \text{ square millimeters (mm}^2) = 1 \text{ square centimeter (cm}^2)$$
$$10,000 \text{ square centimeters} = 1 \text{ square meter (m}^2)$$

Volume
$$1000 \text{ cubic millimeters (mm}^3) = 1 \text{ cubic centimeter (cm}^3)$$
$$1,000,000 \text{ cubic centimeters} = 1 \text{ cubic meter (m}^3)$$

Liquid Capacity
$$1000 \text{ milliliters (mL)} = 1 \text{ liter (L)}$$
$$1000 \text{ cubic centimeters} = 1 \text{ liter}$$

Mass
$$1000 \text{ milligrams (mg)} = 1 \text{ gram (g)}$$
$$1000 \text{ grams} = 1 \text{ kilogram (kg)}$$

Temperature in degrees Celsius (°C)
$$0°C = \text{freezing point of water}$$
$$100°C = \text{boiling point of water}$$

United States Customary Units

Length
$$12 \text{ inches (in.)} = 1 \text{ foot (ft)}$$
$$\left.\begin{array}{r}36 \text{ inches}\\3 \text{ feet}\end{array}\right\} = 1 \text{ yard (yd)}$$
$$\left.\begin{array}{r}5280 \text{ feet}\\1760 \text{ yards}\end{array}\right\} = 1 \text{ mile (mi)}$$

Area
$$144 \text{ square inches (in.}^2) = 1 \text{ square foot (ft}^2)$$
$$9 \text{ square feet} = 1 \text{ square yard (yd}^2)$$

Volume
$$1728 \text{ cubic inches (in.}^3) = 1 \text{ cubic foot (ft}^3)$$
$$27 \text{ cubic feet} = 1 \text{ cubic yard (yd}^3)$$

Liquid Capacity
$$16 \text{ fluid ounces (fl oz)} = 1 \text{ pint (pt)}$$
$$2 \text{ pints} = 1 \text{ quard (qt)}$$
$$4 \text{ quarts} = 1 \text{ gallon (gal)}$$

Weight
$$16 \text{ ounces (oz)} = 1 \text{ pound (lb)}$$

Temperature in degrees Fahrenheit (°F)
$$32°F = \text{freezing point of water}$$
$$212°F = \text{boiling point of water}$$

Time

$$60 \text{ seconds (s)} = 1 \text{ minute (min)}$$
$$60 \text{ minutes} = 1 \text{ hour (h)}$$

1 Introduction to Algebra

1-1 *Variables*

Objective: To simplify numerical expressions and evaluate variable expressions.

Vocabulary

Variable A symbol, such as the letter x, used to represent one or more numbers.

Value of a variable A number that a variable may represent.

Variable expression An expression, such as $x + 5$, that contains a variable.

Numerical expression, or **numeral** An expression, such as $6 + 5$, that names a particular number, called the **value of the expression.**

Simplifying an expression Replacing a numerical expression by the simplest name of its value.

Evaluating an expression Replacing each variable by a given value and simplifying the result.

Symbols   = (is equal to)   ≠ (is not equal to) ab or $a \cdot b$ (multiply)

CAUTION To evaluate ab when $a = 4$ and $b = 5$, be sure to write a multiplication symbol: $ab = 4 \cdot 5$.

Example 1   Simplify: **a.** $4 + (10 - 3)$ **b.** $(20 \div 4) + (6 \div 3)$

Solution   Simplify the numerical expression(s) within parentheses first.

a. $4 + \underbrace{(10 - 3)}$ **b.** $\underbrace{(20 \div 4)} + \underbrace{(6 \div 3)}$
$\underbrace{4 + \quad 7}$ $\underbrace{\quad 5 \quad + \quad 2}$
$\quad 11$ $\quad 7$

Simplify each expression.

1. $5 + (15 - 3)$ **2.** $8 + (30 + 2)$ **3.** $16 - (0 \times 7)$

4. $13 - (4 \times 2)$ **5.** $(9 + 6) + 4$ **6.** $9 + (6 + 4)$

7. $(22 - 17) \times 6$ **8.** $(36 \div 9) + 16$ **9.** $(18 \div 3) + (10 \div 5)$

10. $(8 \times 11) - (3 \times 11)$ **11.** $(24 + 6) \div (30 \div 2)$ **12.** $(42 \div 7) \div (1 \times 3)$

Example 2   Evaluate each expression if $x = 2$ and $y = 3$.

a. xy **b.** $5xy$ **c.** $(6x) - 3$

Solution   Replace x with 2. Replace y with 3. Insert the multiplication symbol(s).

a. $xy = 2 \cdot 3 = 6$ **b.** $5xy = 5 \cdot 2 \cdot 3 = 30$ **c.** $(6x) - 3 = \underbrace{6 \cdot 2} - 3$
$= \underbrace{12 - 3}$
$= \quad 9$

-1   Variables (continued)

aluate each expression if $x = 3$, $y = 5$, and $z = 0$.

. xy	**14.** yz	**15.** xz	**16.** $4xy$
. $6yz$	**18.** $3xz$	**19.** $2xy$	**20.** $7xz$
. $(8x) - 6$	**22.** $(5z) + 4$	**23.** $(2y) - 3$	**24.** $(4x) + 8$

Example 3   Evaluate each expression if $x = 6$, $y = 7$, and $z = 4$.

 a. $(5x) - (2y)$ **b.** $6 \cdot (x + z)$ **c.** $\dfrac{(y + z)}{(y - x)}$

Solution   **a.** $(5x) \quad - \quad (2y)$ $\Big\{$ Replace x with 6 and y with 7 and
 insert the multiplication symbols.

$\underbrace{(5 \cdot 6)}_{30} - \underbrace{(2 \cdot 7)}_{14}$ Simplify the expressions within parentheses.

 16 Subtract.

b. $6 \cdot (x + z)$

$6 \cdot \underbrace{(6 + 4)}_{10}$ Replace x with 6 and z with 4.

$\underbrace{6 \cdot \quad 10}_{60}$ Simplify the expression within parentheses.

 60 Multiply.

c. $\dfrac{(y + z)}{(y - x)} = \dfrac{7 + 4}{7 - 6}$ Replace x with 6, y with 7, and z with 4.

 $= \dfrac{11}{1}$ Simplify the numerator and denominator.

 $= 11$ Divide.

---

aluate each expression if $x = 3$, $y = 5$, and $z = 0$.

$(2y) - (3x)$	**26.** $(3y) + (5x)$	**27.** $(5y) - (6z)$	**28.** $(7z) + (4y)$
$5 \cdot (x - z)$	**30.** $4 \cdot (x + y)$	**31.** $\dfrac{(x + y)}{(y - x)}$	**32.** $\dfrac{(y + z)}{(y + z)}$

xed Review Exercises

form the indicated operations.

0.2×1.3	**2.** $16.35 + 16.07$	**3.** $2.4 \div 0.6$	**4.** $7.3 - 5.6$
$106.4 + 7.8$	**6.** $6.72 - 3.9$	**7.** 50.26×1.2	**8.** $64 \div 0.2$
$\dfrac{3}{4} + \dfrac{1}{10}$	**10.** $\dfrac{3}{8} \times \dfrac{16}{27}$	**11.** $\dfrac{1}{4} \div \dfrac{3}{8}$	**12.** $\dfrac{9}{16} - \dfrac{3}{8}$
$\dfrac{2}{5} \times \dfrac{10}{3}$	**14.** $\dfrac{7}{20} \times \dfrac{5}{21}$	**15.** $\dfrac{5}{8} - \dfrac{1}{2}$	**16.** $\dfrac{6}{7} \div \dfrac{3}{14}$

1–2 Grouping Symbols

Objective: To simplify expressions with and without grouping symbols.

Vocabulary/Symbols

Grouping symbol A symbol used to enclose an expression that should be simplified first. Multiplication symbols are often left out of expressions with grouping symbols. For example:

Parentheses	Brackets	Fraction Bar
$6(5 - 3) = 6 \cdot 2$	$6[5 - 3] = 6 \cdot 2$	$\dfrac{10 + 6}{9 - 5} = \dfrac{16}{4}$

CAUTION When there are no grouping symbols, simplify in the following order:
1. Do all multiplications and divisions in order from left to right.
2. Do all additions and subtractions in order from left to right.

Example 1 Simplify: **a.** $8(7 - 2)$ **b.** $8(7) - 2$

Solution   **a.** $8(7 - 2)$ The parentheses tell you to simplify $7 - 2$ first.

$\qquad\quad$ $8(5)$ $\quad$ $8(5)$ means $8 \cdot 5$.

$\qquad\quad$ 40

$\qquad$ **b.** $8(7) - 2$ Do the multiplication $8 \cdot 7$ first.

$\qquad\quad$ $56 - 2$ Then subtract 2.

$\qquad\quad$ 54

Simplify each expression.

1. a. $9(6 - 1)$ $\qquad$ **2. a.** $12(5 - 3)$ $\qquad$ **3. a.** $6 + 4 \cdot 5$

$\quad$ **b.** $9(6) - 1$ $\qquad\quad$ **b.** $12(5) - 3$ $\qquad\quad$ **b.** $(6 + 4) \cdot 5$

4. a. $8 + 5 \cdot 2$ $\qquad$ **5. a.** $9 - 6 \div 3$ $\qquad$ **6. a.** $12 + 8 \div 4$

$\quad$ **b.** $(8 + 5) \cdot 2$ $\qquad\quad$ **b.** $(9 - 6) \div 3$ $\qquad\quad$ **b.** $(12 + 8) \div 4$

Example 2 Simplify: **a.** $\dfrac{15 + 3}{9 - 3}$ **b.** $\dfrac{8 \cdot 5 + 2}{2(8 - 5)}$

Solution   **a.** $\dfrac{15 + 3}{9 - 3} = \dfrac{18}{6}$ Simplify the numerator and denominator first.

$\qquad\qquad\qquad = 3$ Then divide by 6.

$\qquad$ **b.** $\dfrac{8 \cdot 5 + 2}{2(8 - 5)} = \dfrac{40 + 2}{2(3)}$ Start to simplify the numerator and denominator.

$\qquad\qquad\qquad = \dfrac{42}{6}$ Further simplify the numerator and denominator.

$\qquad\qquad\qquad = 7$ Then divide by 6.

–2 Grouping Symbols (continued)

implify each expression.

7. $\dfrac{6 + 9}{7 - 2}$
    8. $\dfrac{11 - 3}{2 + 6}$
    9. $\dfrac{15 + 3 \cdot 3}{7 + 5}$
    10. $\dfrac{8 \cdot 5 - 4}{3(5 - 3)}$

1. $\dfrac{3(11 - 7)}{2 \cdot 5 - 4}$
    12. $\dfrac{6(13 - 3)}{2 \cdot 5 + 2}$
    13. $\dfrac{6 \cdot 3 + 2 \cdot 7}{2(9 - 5)}$
    14. $\dfrac{7 \cdot 4 - 2 \cdot 5}{3(5 - 3)}$

Example 3   Evaluate each expression if $a = 6$, $b = 2$, $c = 3$, and $d = 0$.

  a. $a(b + c)$ **b.** $\dfrac{8(c + d)}{a - b}$

Solution   **a.** $a(b + c) = 6(2 + 3)$ Replace a with 6, b with 2, and c with 3.
              $= 6(5)$ Simplify the expression within parentheses.
              $= 30$ Multiply.

  b. $\dfrac{8(c + d)}{a - b} = \dfrac{8(3 + 0)}{6 - 2}$ Replace the variables with their given values.

              $= \dfrac{8(3)}{4}$ Simplify the numerator and denominator.

              $= \dfrac{24}{4}$ Divide.

              $= 6$

---

valuate each expression if $x = 2$, $y = 4$, $z = 6$, and $b = 5$.

5. a. $2x + 5$
   b. $2(x + 5)$
    16. a. $5y - 1$
       b. $5(y - 1)$
    17. a. $16 - 3b$
       b. $(16 - 3)b$
    18. a. $3z + 4$
       b. $3(z + 4)$

. a. $bx + y$
  b. $b(x + y)$
    20. a. $xz - b$
       b. $x(z - b)$
    21. a. $2xy + z$
       b. $2(xy + z)$
    22. a. $6xyz - b$
       b. $6x(yz - b)$

**. ** $5(4y - 3x)$
    24. $6z - 2xy$
    25. $xyz - 5$
    26. $x(y \cdot y + z)$

**. ** $\dfrac{9x + z}{x + z}$
    28. $\dfrac{8x - z}{z - b}$
    29. $\dfrac{9y - z}{5(b - y)}$
    30. $\dfrac{2(x + y)}{x + y}$

lixed Review Exercises

mplify.

**. ** $(12 - 6) \div 3$
    2. $20 \cdot 8 + 18 \cdot 2$
    3. $5 \times (25 - 7)$

**. ** $9 + 15 \div 3$
    5. $(25 + 3) \div (8 \div 2)$
    6. $(7 + 5) \cdot (8 - 2)$

valuate each expression if $a = 2$, $b = 3$, and $c = 4$.

**. ** $5ab$
    8. bc
    9. $(2c) - 3$

**. ** $\dfrac{a + c}{c - a}$
    11. $(7a) - (4b)$
    12. $6a$

1–3 Equations

Objective: To find solution sets of equations over a given domain.

Vocabulary

Equation An equation is formed by placing an equals sign between two numerical or variable expressions. Examples: $2 + 3 = 5$, $x - 1 = 7$

Open sentence A sentence containing variables.

Domain of a variable The given set of numbers a variable may represent.

Solution, or root, of an equation A value of a variable that turns an open sentence into a true statement. For example, 8 is the solution of the equation $x - 1 = 7$.

Solution set of an equation The set of all the solutions of an equation.

Symbol

$\in$ (is an element of, or belongs to)

Example 1 The domain of x is $\{0, 1, 2\}$.
Is the equation $3x - 1 = 5$ true when $x = 0$? when $x = 1$? when $x = 2$?

Solution Replace x in turn by 0, 1, and 2.

x	$3x - 1 = 5$	
0	$3 \cdot 0 - 1 = 5$	False
1	$3 \cdot 1 - 1 = 5$	False
2	$3 \cdot 2 - 1 = 5$	**True**

Example 2 Read: **a.** $y \in \{1, 2, 3\}$ **b.** $x \in \{0, 2, 4, 6\}$

Solution **a.** y belongs to the set whose members are 1, 2, and 3.

b. x belongs to the set whose members are 0, 2, 4, and 6.

Example 3 Solve $y(3 - y) = 2$ if $y \in \{0, 1, 2, 3\}$.

Solution Replace y in turn with 0, 1, 2, and 3.

y	$y(3 - y) = 2$	
0	$0(3 - 0) = 2$	False
1	$1(3 - 1) = 2$	**True**
2	$2(3 - 2) = 2$	**True**
3	$3(3 - 3) = 2$	False

The solutions are 1 and 2.

The solution set is $\{1, 2\}$.

-3 Equations (continued)

lve each equation if $x \in \{0, 1, 2, 3, 4, 5\}$.

. $x + 3 = 7$ **2.** $5 + x = 9$ **3.** $x - 3 = 2$

. $x - 1 = 3$ **5.** $6 - x = 1$ **6.** $5 - x = 2$

. $x + 3 = 3$ **8.** $x + 2 = 5$ **9.** $2x = 8$

. $3x = 12$ **11.** $4x = 0$ **12.** $5x = 25$

. $x \div 2 = 1$ **14.** $x \div 1 = 3$ **15.** $\frac{1}{2}x = 1$

. $4x = 16$ **17.** $x \cdot x = 4$ **18.** $5x = 5$

. $x \cdot x = 9$ **20.** $x \cdot x = 16$ **21.** $3x + 7 = 19$

. $5x - 2 = 13$ **23.** $x(5 - x) = 6$ **24.** $x(4 - x) = 3$

Example 4   Solve over the domain $\{2, 4, 6\}$.
Three more than twice a number is 11. What is the number?

Solution   Use mental math to see which members of the given domain are solutions.

Number	Three more than twice a number is 11.	
2	Three more than twice 2 is 11.	False
4	Three more than twice 4 is 11.	**True**
6	Three more than twice 6 is 11.	False

The number is 4.

---

lve each problem over the domain $\{2, 3, 4, 5\}$.

. Eleven more than a number is 15. What is the number?

. Four times a number is 16. What is the number?

. A number divided by one is 5. What is the number?

. Two less than a number is 3. What is the number?

. One less than twice a number is 9. What is the number?

. One more than twice a number is 7. What is the number?

ixed Review Exercises

nplify.

. $9 \cdot 8 + 9 \cdot 12$ **2.** $8 + (12 \div 2)$ **3.** $(16 - 7) \div 3$

. $(2 + 3 \cdot 4) \div 7$ **5.** $15 - 9 \div 3 \div 3$ **6.** $35 \div 7 \div (3 + 2)$

aluate if $a = 2$, $x = 3$, $y = 5$, and $z = 6$.

. $2x + 3y$ **8.** $8 \cdot (z - x)$ **9.** $3(ax + 2)$

. $3xz + 2y$ **11.** $axz \div (y + 1)$ **12.** $4a \div (x + 1)$

NAME _____ DATE _____

1-4 Translating Words into Symbols

Objective: To translate phrases into variable expressions.

	Phrase	Variable Expression
Addition	The *sum* of 6 and x A number *increased* by 5 3 *more than* a number	$6 + x$ $n + 5$ $n + 3$
Subtraction	The *difference* between a number and 7 A number *decreased* by 6 3 *less than* a number 9 *minus* a number	$x - 7$ $x - 6$ $n - 3$ $9 - n$
Multiplication	The *product* of 6 and a number Five *times* a number One half *of* a number	$6n$ $5n$ $\frac{1}{2}x$
Division	The *quotient of* a number and 4 A number *divided* by 8	$\frac{n}{4}$ $\frac{n}{8}$

CAUTION The phrase "6 less than x" is translated $x - 6$ and *not* $6 - x$.
The phrase "6 more than x" can be translated as either $6 + x$ or $x + 6$.

Example 1 Translate each phrase into a variable expression.

    a. Five less than half of x **b.** One half the difference between x and 5

Solution **a.** Half of x: $\frac{1}{2}x$ **b.** The difference between x and 5: $(x - 5)$

        Five less than half of x: $\frac{1}{2}x - 5$ One half the difference between x and 5: $\frac{1}{2}(x - 5)$

Translate each phrase into a variable expression. Use n for the variable.

1. Five more than a number

2. The product of 7 and a number

3. A number divided by 4

4. Seven less than a number

5. The sum of 3 and a number

6. A number decreased by 8

7. The quotient of a number and 3

8. The difference between a number and 6

9. Nine times a number

10. The difference between 9 and a number

11. Four more than half a number

12. Two less than 4 times a number

13. Ten less than one half a number

14. The quotient of 3 and a number

15. Three plus the product of a number and 5

16. The difference between 3 times a number and

17. Four more than three times a number

18. Nine increased by twice a number

19. Eight times the sum of a number and 2

20. Seven times the difference of a number and 5

–4 Translating Words into Symbols *(continued)*

Example 2	Complete the statement with a variable expression.	**Solution**
	Lila is 3 in. shorter than Dale.	
	If Dale's height is x in., then Lila's height is __?__ in.	$x - 3$

Complete each statement with a variable expression.

1. Lisa is 2 cm taller than Fred.
   If Fred's height is f cm, then Lisa's height is __?__ cm.

2. Arnie is 7 in. shorter than Rick.
   If Rick's height is r in., then Arnie's height is __?__ in.

3. Shawn has $6 more than Maria.
   If Maria has m dollars, then Shawn has __?__ dollars.

4. Barb has twice as much money as Carlos.
   If Carlos has c dollars, then Barb has __?__ dollars.

Example 3	Complete each statement with a variable expression.	**Solution**
	a. The sum of two numbers is 15.	**a.** $15 - x$
	If one number is x, then the other number is __?__.	
	b. The product of two numbers is 24.	**b.** $24 \div y$, or $\dfrac{24}{y}$
	If one number is y, then the other number is __?__.	

Complete each statement with a variable expression.

25. The sum of two numbers is 9.
    If one number is n, then the other
    number is __?__.

26. The product of two numbers is 15.
    If one number is y, then the other
    number is __?__.

27. The sum of two numbers is 12.
    If one number is x, then the other
    number is __?__.

28. The product of two numbers is 20.
    If one number is w, then the other
    number is __?__.

Mixed Review Exercises

Evaluate if $t = 3$, $x = 4$, $y = 5$, and $z = 6$.

1. $5x - 2$ 2. $2 + xyz$ 3. $(3x - 1) \cdot 2$

4. $2x + 3y - t$ 5. $tyz - 1$ 6. $xy + t + z$

Solve if $x \in \{0, 1, 2, 3, 4\}$.

7. $x + 3 = 7$ 8. $3x = 6$ 9. $5x = 5$ 10. $3x = 0$

11. $2x - 1 = 5$ 12. $5 = 2x + 1$ 13. $2x = x + 1$ 14. $x \div 4 = 1$

1–5 Translating Sentences into Equations

Objective: To translate word sentences into equations.

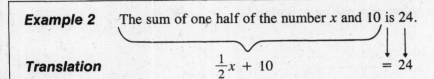

Example 1	Twice the sum of a number and 3 is twelve.
Translation	$2 \cdot \quad (n + 3) \quad = \quad 12$

Example 2	The sum of one half of the number x and 10 is 24.
Translation	$\frac{1}{2}x + 10 \quad = 24$

Match the sentence in the first column with the corresponding equation in the second column.

1. Three more than twice a number is nine.

2. Two less than three times a number is nine.

3. Three times the number which is two less than x is nine.

4. Two times the number which is three less than x is nine.

5. Two times the quantity three more than x is nine.

6. Three less than the product of two and x is nine.

7. Two decreased by three times a number is nine.

8. Three times the quantity two decreased by x is nine.

a. $2 - 3x = 9$

b. $3(x - 2) = 9$

c. $2x + 3 = 9$

d. $2(x + 3) = 9$

e. $3(2 - x) = 9$

f. $2(x - 3) = 9$

g. $2x - 3 = 9$

h. $3x - 2 = 9$

Translate each sentence into an equation.

9. One half of a number is four.

10. Three more than a number is eight.

11. Six less than a number is nine.

12. Two less than three times a number is eleven.

13. Twice a number is 12 more than five times the number.

14. The number x is seven more than one fourth of itself.

15. Five less than twice a number is 15.

16. Two times the quantity x minus 1 is 12.

17. Eleven more than twice x is five less than x.

18. Nine times x is twice the sum of x and five.

Vocabulary

Formulas Equations that state rules about relationships. Examples:

$A = lw$	**Area of rectangle** = length of rectangle × width of rectangle
$P = 2l + 2w$	**Perimeter of rectangle** = (2 × length) + (2 × width)
$D = rt$	**Distance traveled** = rate × time traveled
$C = np$	**Cost** = number of items × price per item

–5 Translating Sentences into Equations (continued)

Example 3 Use the figure and the information below it to write an equation involving x.

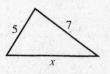

Solution Perimeter = the sum of the lengths of the sides.
$$20 = 5 + 7 + x$$
$$20 = 12 + x$$

Perimeter = 20

se the figure and the information below to write an equation
volving x.

x 6
7
Perimeter = 21

20.

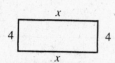

x
4   4
x
Perimeter = 28

21.

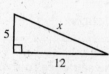

5   x
12
Perimeter = 30

Example 4 **a.** Choose a variable to represent the number described in parentheses.
b. Write an equation that represents the given information.

The distance traveled in 4 h of driving was 260 km. (Hourly rate)

Solution 1 **a.** Let r = the hourly rate
b. Rate × time = distance
$$r \cdot 4 = 260,$$
$$\text{or } 4r = 260.$$

Solution 2 **a.** Let r = the hourly rate
b. Since the hourly rate is the number of km traveled in one hour, $r = \dfrac{260}{4}$.

Exercises 22–24,
Choose a variable to represent the number described in parentheses.
Write an equation that represents the given information.

The distance traveled in 3 h of driving was 210 km. (Hourly rate)

A train traveled at 66 km/h for 4 h. (Distance traveled)

A driver averaged 60 km/h while driving 300 km. (Time)

xed Review Exercises

e if $x \in \{0, 1, 2, 3, 4, 5, 6\}$.

$3 + x = 8$ **2.** $4 = x - 2$ **3.** $4x = 20$ **4.** $2 = x \div 3$

$2x + 1 = 7$ **6.** $3x = x + 4$ **7.** $x + 3 = 2x$ **8.** $2x = x \cdot 2$

nslate each phrase into a variable expression.

A number increased by 6   **10.** The quotient of x and 2

The product of 9 and a number   **12.** Twice the sum of a number and 3

NAME _____ DATE _____

1–6 Translating Problems into Equations

Objective: To translate word problems into equations.

Step 1 **Read the problem carefully.**

Step 2 **Choose a variable and represent the unknowns.**

Step 3 **Reread the problem and write an equation.**

Example 1 Translate the problem into an equation. Do not solve the equation.

(1) Rudy has $15 more than Franco.
(2) Together they have $65.
How much money does each have?

Solution Use the three steps shown above.

Step 1 The unknowns are the amounts of money Rudy and Franco have.
Each numbered sentence gives you a fact.

Step 2 Choose a variable for one unknown: Let f = Franco's amount.
Use f and fact (1):   Then $f + 15$ = Rudy's amount

Step 3 Use f and fact (2) to write an equation: $f + (f + 15) = 65$.

Translate each problem into an equation. Do not solve the equation.

1. (1) Shelley has $10 more than Michele.
   (2) Together they have $50.
   How much money does each have?

2. (1) Bart has twice as much money as Elmer.
   (2) Together they have $75.
   How much money does each have?

3. (1) Sandy sold four more cars than Michael.
   (2) Together they sold a total of 14 cars.
   How many cars did each sell?

4. (1) Joseph worked 8 h less than Lois.
   (2) Together they worked 72 h.
   How many hours did each work?

5. (1) Gloria spent $2 more for dinner than Lucy.
   (2) Together they spent $26.
   How much did each spend?

6. (1) Nancy jogged 3 mi less than Pat.
   (2) Together they jogged 7 mi.
   How far did each jog?

7. (1) Ray has three times as many compact discs as Mia.
   (2) Together they have 20 compact discs.
   How many compact discs does each have?

1–6 *Translating Problems into Equations* (continued)

Example 2   Translate the problem into an equation. Do not solve the equation.

(1) A 12 ft piece of pipe is cut into two pieces.
(2) One piece is 2 ft longer than the other.
What are the lengths of the pieces?

Solution   Use the three steps shown on page 11.
Make a sketch to help you understand the problem.

Step 1   The unknowns are the lengths of the pieces.
Sentences (1) and (2) each give a fact.

Step 2   Choose a variable for one unknown:
Let x = the shorter piece.

Use x and sentence (2):
Then $x + 2$ = the longer piece.

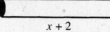

Step 3   Use fact (1) to write an equation: $x + (x + 2) = 12$.

Translate each problem into an equation. Draw a sketch to help you.

8. (1) A 25 ft piece of cable is cut into two pieces.
(2) One piece is 5 ft longer than the other.
What are the lengths of the pieces?

9. (1) Kim walked twice as far as Paula.
(2) The sum of the distances they walked is 12 km.
How far did each walk?

0. (1) Miguel drove 20 km more than Jorge.
(2) Miguel drove three times as far as Jorge.
How far did each drive?

1. (1) A 16 ft piece of wood is cut into 2 pieces.
(2) One piece is 4 ft shorter than the other.
How long is each piece?

Mixed Review Exercises

Solve if $x \in \{0, 1, 2, 3, 4, 5, 6\}$.

1. $2x + 1 = 7$ **2.** $x \div 3 = 2$ **3.** $15 = 15x$ **4.** $5 = 3x - 1$

5. $3 + 3x = 15$ **6.** $5 = 2x - 7$ **7.** $4x = x + 3$ **8.** $x \cdot x = 9$

Translate each phrase into a variable expression.

9. One half of a number

10. Two more than five times a number

1. Five less than three times a number

12. Four more than one third a number

1-7 A Problem Solving Plan

Objective: To use the five-step plan to solve word problems over a given domain.

Plan for Solving a Word Problem

Step 1 Read the problem carefully.
Find what the unknowns are and what the facts are.
Making a sketch may help.

Step 2 Choose a variable.
Use it with the given facts to represent the unknowns.

Step 3 Reread the problem.
Write an equation that represents relationships among the numbers.

Step 4 Solve the equation and find the unknowns.

Step 5 Check your results with the words of the problem.
Give the answer.

Symbol $\stackrel{?}{=}$ (Are they equal?)

Example   Solve using the five-step plan. Write out each step. A choice of possible numbers for one unknown is given.

One number is 12 more than another number. Their sum is 108.
Find the numbers.
Choices for the smaller number: 40, 48, 56

Solution

Step 1   The unknowns are the two numbers.

Step 2   Let n = the smaller number. Then $n + 12$ = the larger number.

Step 3 $n + (n + 12) = 108$.

Step 4   Replace n in turn by 40, 48, and 56.

n	$n + (n + 12) = 108$	
40	$40 + (40 + 12) \stackrel{?}{=} 108$	False
48	$48 + (48 + 12) \stackrel{?}{=} 108$	**True**
56	$56 + (56 + 12) \stackrel{?}{=} 108$	False

Smaller number: $n = 48$
Larger number: $n + 12 = 60$

Step 5   Check the results of Step 4 with the words of the problem.

One number is 12 more than another number. $60 - 48 \stackrel{?}{=} 12$
$12 = 12 \checkmark$

Their sum is 108. $48 + 60 \stackrel{?}{=} 108$
$108 = 108 \checkmark$

The numbers are 48 and 60.

–7 A Problem Solving Plan (continued)

Solve using the five-step plan. Write out each step. A choice of possible numbers for one unknown is given.

1. In the chorus there are 12 more girls than boys. There are 32 students in the chorus. How many are boys? How many are girls?
   Choices for the number of boys: 8, 10, 12

2. There were twice as many four-door cars as two-door cars in the parking lot. There were 36 cars in the lot. How many had four doors? How many had two doors?
   Choices for the number of four-door cars: 12, 18, 24

3. One number is 52 more than another number. Their sum is 176. Find the numbers.
   Choices for the smaller number: 62, 72, 82

4. One number is three times another number. Their sum is 180. Find the numbers.
   Choices for the smaller number: 25, 35, 45

5. Art has three times as much money as Flora. Together they have $180. How much money does each person have?
   Choices for Flora's amount: 45, 60, 75

6. In a class election, the winner had 18 more votes than the loser. If 104 class members voted, how many votes did the winner receive? How many votes did the loser receive?
   Choices for the loser's votes: 29, 43, 61

7. One number is twice another number. The larger number is also 32 more than the smaller number. Find the numbers.
   Choices for the smaller number: 32, 40, 48

8. The house is 12 years older than the garage. The house is also three times as old as the garage. How old is each building?
   Choices for the garage's age: 3, 6, 9

9. Lupita has five times as much money as David. Lupita also has $20 more than David. How much money does each person have?
   Choices for David's amount: 4, 5, 6

Mixed Review Exercises

Simplify.

1. $\dfrac{6 \cdot 3 + 6 \cdot 7}{13 - 5 + 12}$

2. $(27 - 3 + 6 \div 2) \div (5 + 4)$

3. $50 - (16 \div 4 + 2)$

4. $8 \cdot 15 + 15 \cdot 2$

Translate each sentence into an equation.

5. Five times a number is 30.

6. Five is three less than a number.

7. One more than twice a number is 15.

8. One fifth of a number is four.

1–8 Number Lines

Objective: To graph real numbers on a number line and to compare real numbers.

Symbols   … (and so on)   < (is less than)   > (is greater than)

Vocabulary

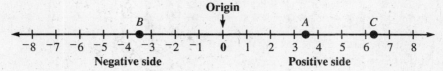

Negative side   Positive side

Graph of a number C is the graph of $6\frac{1}{3}$.   **Coordinate of a point** $6\frac{1}{3}$ is the coordinate of C

Positive integers $\{1, 2, 3, \ldots\}$ **Positive number** Example: 3.5

Negative integers $\{^-1, \,^-2, \,^-3, \ldots\}$ **Negative number** Example: $^-3.5$

Integers $\{\ldots, \,^-3, \,^-2, \,^-1, 0, 1, 2, 3, \ldots\}$ **Whole numbers** $\{0, 1, 2, 3, \ldots\}$

Real number Any number that is either positive, negative, or zero.

Example 1   Write a number to represent each situation.

    a. A deposit of \$5: 5   **b.** 6.2 km east: 6.2   **c.** 10° above freezing: 10
       A withdrawal of \$5: $^-5$ 6.2 km west: $^-6.2$ 10° below freezing: $^-10$

**Write a number to represent each situation. Then write the opposite of
that situation and write a number to represent it.**

1. Five wins
2. Six floors up
3. A gain of eight yards
4. Three points under par
5. 150 m above sea level
6. 22 km east
7. Six bonus points
8. Ten steps down
9. Four steps to the right
10. 7° above freezing (0 °C)
11. A gain of two pounds
12. Latitude of 20° north
13. A bank deposit of \$50
14. A loss of \$24

Example 2   Translate each statement into symbols.

    a. Three is greater than negative five   **b.** Negative six is less than negative two.

Solution   **a.**   3   > $^-5$ **b.** $^-6$ < $^-2$

Translate each statement into symbols.

15. Four is greater than negative two.
16. Negative ten is less than negative two.
17. Negative four is greater than negative six.
18. Seven is less than nine.
19. Two is less than two and two tenths.
20. Negative five tenths is less than zero.
21. Negative 12 is less than three.
22. One half is greater than one fifth.

Study Guide, ALGEBRA, Structure and Method, Book 1

15

1–8 Number Lines (continued)

Example 3 List the letters of the points on the number line whose coordinates are given:
$^-4, 2.5, 0$

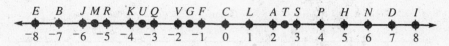

Solution K, T, C

List the letters of the points whose coordinates are given. Use the number line in Example 3.

23. $^-6, 3$ 24. $^-3, 5$ 25. $0, ^-7$ 26. $^-5, 1$

27. $2, ^-2, ^-3\frac{1}{2}$ 28. $0, 1, ^-1\frac{1}{2}$ 29. $^-3, ^-5\frac{1}{2}, ^-2$ 30. $^-7, ^-5, 2\frac{1}{2}$

Example 4 State two inequalities, one with > and one with < , relating the coordinates of the points indicated by the heavy dots.

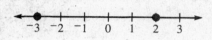

Solution On the number line, a number to the right of another is the greater. $2 > ^-3,\ ^-3 < 2$

State two inequalities, one with > and one with < , relating the coordinates of the points indicated by the heavy dots.

31. 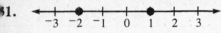   32.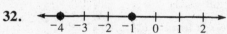

33. 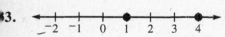   34.

35. 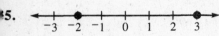   36.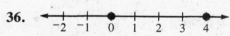

37. 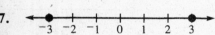   38.

Mixed Review Exercises

Evaluate if $a = 2$, $b = 4$, $c = 3$, $x = 5$, and $y = 6$.

1. $2ab - 3x$ 2. $3a(y - b)$ 3. $x(2y \div 6)$

4. $\frac{1}{2}(5x - 3)$ 5. $4y - (5b \div x)$ 6. $\left(\frac{1}{3}ab\right) + (5 - x) \cdot c$

Translate each sentence into an equation.

7. Nine more than a number is five.   8. The product of a number and 9 is 27.

9. One third of a number is four.   10. A number decreased by 18 is 30.

1–9 *Opposites and Absolute Values*

Objective: To use opposites and absolute values

Vocabulary

Opposite of a number Each number in a pair such as 3 and ⁻3 is called
the opposite of the other number. The opposite of 0 is 0.

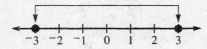

Absolute value of a number The distance between the graph of the number
and 0 on a number line. For example, the absolute value of both 3 and
⁻3 is the positive number 3.

Symbols

The *opposite of a* is written $-a$. Since -3 and ⁻3 name the same number,
you can use the lowered minus sign to write negative numbers from now on.

$|a|$ means *the absolute value of a.*

Examples: $|3| = 3$ $|-3| = 3$ $|0| = 0$

CAUTION $-a$, read "the opposite of a," is not necessarily a negative
number. For example, if $a = -3$, then $-a = -(-3) = 3$.

Example 1	Simplify:	**a.** $-(5 + 4)$	**b.** $-(-2.1)$	**c.** $-(8 - 5)$
Solution		**a.** $-(9) = -9$	**b.** 2.1	**c.** $-(3) = -3$

Simplify.

1. $-(-3)$ 2. $-(9)$ 3. $-(2 + 6)$ 4. $-(6 - 2)$ 5. $-(5 + 9)$

6. $-(6 - 6)$ 7. $-(6 + 6)$ 8. $-(-3.5)$ 9. $-(2.9)$ 10. $-(7 - 3)$

Example 2	Simplify:	**a.** $|-2| + 3$	**b.** $|-3| - |-2|$
Solution		**a.** $|-2| + 3 = 2 + 3 = 5$	**b.** $|-3| - |-2| = 3 - 2 = 1$

Simplify.

11. $6 + |-4|$ 12. $|-9| + 2$ 13. $|-2| + |-6|$

14. $|-1.5| + |1.5|$ 15. $|-0.2| + |-1.8|$ 16. $\left|-\frac{1}{4}\right| + 0$

17. $\left|-\frac{3}{4}\right| + \left|\frac{1}{4}\right|$ 18. $|-6| + |-0.5|$ 19. $|8| - |-6|$

20. $\left|-\frac{4}{5}\right| - \left|\frac{1}{5}\right|$ 21. $|-8| - |8|$ 22. $|7| - |-7|$

–9 Opposites and Absolute Values (continued)

Example 3   Use one of the symbols $>$, $<$, or $=$ to make a true statement.

      a. $-8 \underline{\ ?\ } -(-8)$ **b.** $|-6| \underline{\ ?\ } 6$

Solution   **a.** $-8 \underline{\ ?\ } 8$ **b.** $6 \underline{\ ?\ } 6$

           $-8 < 8$ $6 = 6$

---

se one of the symbols $>$, $<$, or $=$ to make a true statement.

3. $-(-3) \underline{\ ?\ } -3$ **24.** $-2 \underline{\ ?\ } -(-2)$ **25.** $-(-4) \underline{\ ?\ } |-5|$

6. $|-8| \underline{\ ?\ } |8|$ **27.** $|-12| \underline{\ ?\ } |-8|$ **28.** $|-6| \underline{\ ?\ } |-10|$

9. $-|-3| \underline{\ ?\ } -3$ **30.** $-2 \underline{\ ?\ } -|-2|$ **31.** $-|-9| \underline{\ ?\ } |-7|$

Example 4   Solve each equation over the set of real numbers. If there is no solution, explain why there is none.

      a. $|x| = 2$ **b.** $|n| = -5$

Solution   **a.** Both -2 and 2 are 2 units from the origin, so the replacements for x that make $|x| = 2$ true are -2 and 2. Therefore the solution set is $\{-2, 2\}$.

      b. The absolute value of a number is never negative. Therefore there is no solution.

---

olve each equation over the set of real numbers. If there is no
olution, explain why there is none.

2. $|n| = 1$ **33.** $|x| = 5$ **34.** $|t| = \dfrac{1}{4}$

5. $|z| = 0.5$ **36.** $|a| = -3$ **37.** $|m| = -6$

8. $|-q| = 7$ **39.** $|-c| = 2$ **40.** $|-w| = -4$

Mixed Review Exercises

implify.

1. $7 - (3 + 2)$ **2.** $7 - 3 + 2$ **3.** $12 \div (2 + 4)$

4. $(7 - 2) \cdot (5 - 3)$ **5.** $7 - 3 \cdot 2$ **6.** $18 - 10 \div (3 + 2)$

Vrite a number to represent each situation. Then write the opposite
f that number.

7. Four steps up

8. A bank deposit of $80

9. A loss of $100

0. Nine kilometers east

Study Guide, ALGEBRA, Structure and Method, Book 1

2 Working with Real Numbers

2-1 *Basic Assumptions*

Objective: To use number properties to simplify expressions.

Vocabulary

Unique One and only one

Terms When a and b are added, a and b are called terms.

Factors When a and b are multiplied, a and b are called factors.

Properties of Real Numbers	Addition	Multiplication
Closure Properties The sum and product of any two real numbers are also real numbers and they are unique.	$2 + 3 = 5$ and only 5	$2 \cdot 3 = 6$ and only
Commutative Properties The order in which you add or multiply any two real numbers does not affect the result.	$3 + 5 = 5 + 3$	$3 \cdot 5 = 5 \cdot 3$
Associative Properties When you add or multiply any three real numbers, the grouping (or association) of the numbers does not affect the result.	$(3 + 4) + 6 = 3 + (4 + 6)$	$(3 \cdot 4)5 = 3(4 \cdot 5)$

Example 1 Simplify: **a.** $75 + 13 + 25 + 47$ **b.** $4 \cdot 7 \cdot 25 \cdot 3$

Solution Regrouping makes mental math easier.

a. $75 + 13 + 25 + 47 = (75 + 25) + (13 + 47)$ Regroup the terms.
$\qquad\qquad\qquad\qquad\quad = 100 + 60$ Simplify within the
$\qquad\qquad\qquad\qquad\quad = 160$ parentheses. Add.

b. $4 \cdot 7 \cdot 25 \cdot 3 = (4 \cdot 25)(7 \cdot 3)$ Regroup the factors.
$\qquad\qquad\qquad\quad = 100 \cdot 21$ Simplify within the parentheses.
$\qquad\qquad\qquad\quad = 2100$ Multiply.

Example 2 Simplify $1\frac{1}{3} + 16\frac{4}{5} + 2\frac{2}{3} + 3\frac{1}{5}$.

Solution Regroup the fractions. Simplify within the parentheses. Add.

$$1\frac{1}{3} + 16\frac{4}{5} + 2\frac{2}{3} + 3\frac{1}{5} = \left(1\frac{1}{3} + 2\frac{2}{3}\right) + \left(16\frac{4}{5} + 3\frac{1}{5}\right)$$
$$= 4 + 20$$
$$= 24$$

–1 Basic Assumptions (continued)

Example 3 Simplify $0.8 + 3.7 + 0.2 + 5.3$.

Solution Regroup the decimals. Simplify within the parentheses. Add.
$$0.8 + 3.7 + 0.2 + 5.3 = (0.8 + 0.2) + (3.7 + 5.3)$$
$$= 1 + 9$$
$$= 10$$

Simplify.

1. $125 + 42 + 75 + 28$ **2.** $507 + 36 + 43 + 14$

3. $2 \cdot 18 \cdot 5 \cdot 4$ **4.** $40 \cdot 3 \cdot 4 \cdot 20$

5. $50 \cdot 27 \cdot 4 \cdot 2$ **6.** $4 \cdot 15 \cdot 25 \cdot 3$

7. $3\frac{1}{2} + 5\frac{2}{3} + 2\frac{1}{2} + \frac{1}{3}$ **8.** $7\frac{2}{3} + 4\frac{3}{5} + 2\frac{1}{3} + \frac{12}{5}$

9. $0.2 + 3.9 + 2.8 + 0.1$ **10.** $0.6 + 5.2 + 0.4 + 3.8$

11. $2.85 + 3.75 + 1.15 + 9.25$ **12.** $3.25 + 1.95 + 8.75 + 11.05$

Example 4 Simplify: **a.** $6 + 8m + 4 + 7n$ **b.** $(3w)(2x)(4y)(5z)$

Solution **a.** $6 + 8m + 4 + 7n = 8m + 7n + (6 + 4)$ Regroup the terms.
         $= 8m + 7n + 10$ Simplify.

     b. $(3w)(2x)(4y)(5z) = (3 \cdot 2 \cdot 4 \cdot 5)(wxyz)$ Regroup the factors.
         $= 120wxyz$ Simplify.

Simplify.

13. $2 + 5y + 8$ **14.** $9 + 5z + 11$ **15.** $4 + 3x + 5$ **16.** $3 + 2w + 4$

17. $3(20a)$ **18.** $4(5n)$ **19.** $(5x)(6y)$ **20.** $(8m)(5n)$

21. $(6x)(y)(4z)$ **22.** $(2p)(3q)(5r)$ **23.** $(3a)(7b)(c)$ **24.** $(e)(6f)(2g)$

25. $a + 2 + b + 5$ **26.** $9 + x + y + 3$ **27.** $3p + 4 + 2q + 6$ **28.** $7m + 1 + 5n + 4$

29. $4 + 6x + 2 + 3y$ **30.** $6p + 3 + 2q + 37$ **31.** $(5a)(4b)(25c)(8d)$ **32.** $(4w)(2x)(5y)(5z)$

Mixed Review Exercises

Evaluate if $a = 2$, $x = 4$, $y = 6$, and $z = 3$.

1. $\dfrac{3x - a}{a + z}$ **2.** $4z(y - a)$ **3.** $\dfrac{2a + x}{3z - (y + 2)}$

Simplify.

4. $|-3| + |-5|$ **5.** $\left|-\dfrac{1}{6}\right| + 0$ **6.** $|-3.2| + |3.2|$

7. $|8| - |-8|$ **8.** $|-4| - |-2|$ **9.** $\left|-\dfrac{5}{7}\right| - \left|\dfrac{3}{7}\right|$

NAME _____ DATE _____

2–2 Addition On A Number Line

Objective: To add real numbers using a number line or properties about opposites.

Properties	Examples
Identity Property of Addition The sum of a real number and 0 is identical to the number itself. $a + 0 = a$ and $0 + a = a$	$5 + 0 = 5$ and $0 + 5 = 5$
Properties of Opposites Every real number has an opposite. The sum of a real number and its opposite is 0. $a + (-a) = 0$ and $(-a) + a = 0$	$3 + (-3) = 0$ and $(-3) + 3 = 0$
Property of the Opposite of a Sum For all real numbers a and b: $-(a + b) = (-a) + (-b)$	$-(2 + 3) = -5 = (-2) + (-3)$

Example 1 Simplify: **a.** $3 + 4$ **b.** $-3 + (-4)$ **c.** $2 + (-5)$ **d.** $-2 + 5$

Solution **a.**

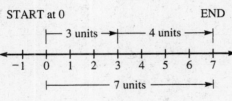

$$3 + 4 = 7$$

b.

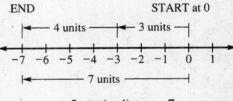

$$-3 + (-4) = -7$$

c.

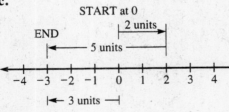

$$2 + (-5) = -3$$

d.

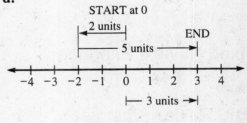

$$-2 + 5 = 3$$

Simplify each expression. If necessary, draw a number line to help you.

1. $4 + 2$

2. $-4 + (-2)$

3. $6 + (-9)$

4. $-6 + 9$

5. $3 + 5$

6. $-3 + (-5)$

7. $-8 + 4$

8. $8 + (-4)$

9. $(-2 + 5) + 4$

10. $(-4 + 7) + 1$

11. $(-3 + 1) + 2$

12. $(-6 + 5) + 3$

2–2 Addition On A Number Line (continued)

Simplify each expression. If necessary, draw a number line to help you.

13. $(-6 + 9) + (-2)$ **14.** $(-4 + 7) + (-3)$

15. $(-3 + 5) + 7$ **16.** $(-8 + 12) + (-6)$

17. $[-6 + (-5)] + 6$ **18.** $[-3 + (-7)] + 3$

19. $25 + [7 + (-2)]$ **20.** $32 + [8 + (-14)]$

21. $[19 + (-9)] + [2 + (-2)]$ **22.** $(-2 + 2) + [26 + (-6)]$

23. $[0 + (-6)] + [-5 + (-25)]$ **24.** $(-7 + 7) + [14 + (-4)]$

25. $-1 + (-2) + (-7)$ **26.** $(-4) + (-6) + (-9)$

27. $-3 + (-11) + 8 + (-5)$ **28.** $-10 + 14 + (-3) + (-12)$

29. $-5.3 + (-1.5) + 6.8$ **30.** $4.2 + (-3.4) + (-6.9)$

31. $-\dfrac{3}{4} + 3 + \left(-\dfrac{9}{4}\right)$ **32.** $-\dfrac{11}{3} + (-4) + \left(-\dfrac{7}{3}\right)$

Example 2   Simplify $3 + (-2) + x + 6$.

Solution   Use the commutative and associative properties to regroup.

$$3 + (-2) + x + 6 = 1 + x + 6$$
$$= x + (1 + 6)$$
$$= x + 7$$

Simplify.

33. $2 + x + (-6)$ **34.** $y + (-1) + 5$

35. $3n + 4 + (-1)$ **36.** $5 + 2n + (-4)$

37. $12 + 4n + (-20)$ **38.** $-8 + (-2z) + 11$

Mixed Review Exercises

Simplify.

1. $6 + 8 \div 2$ **2.** $|-5| + |2|$

3. $9 - [-(-1)]$ **4.** $\dfrac{2 + (3 \cdot 6)}{4}$

5. $\left|\dfrac{3}{5}\right| + \left|-\dfrac{2}{5}\right|$ **6.** $\left|-\dfrac{3}{4}\right| + \left|-\dfrac{1}{4}\right|$

7. $52 + 35 + 48 + 15$ **8.** $12 \cdot 5 \cdot 30$

9. $(2x)(4y)(5z)$ **10.** $4\dfrac{3}{4} + 2\dfrac{1}{3} + 5\dfrac{1}{4} + \dfrac{2}{3}$

11. $0.3 + 2.6 + 5.7$ **12.** $4 \cdot 19 \cdot 25$

2–3 Rules for Addition

Objective: To add real numbers using rules for addition.

Vocabulary

Opposite signs A positive and a negative number are said to have opposite signs.

Rules for Addition	Examples
If two numbers have the *same sign*, add their absolute values and put their common sign before the result.	$2 + 5 = 7$ $-2 + (-5) = -7$
If two numbers have *opposite signs*, subtract the lesser absolute value from the greater and put the sign of the number having the greater absolute value before the result.	$6 + (-4) = 6 - 4 = 2$ $(-6) + 4 = -(6 - 4) = -2$
If two numbers are *opposites*, then their sum is zero.	$3 + (-3) = 0$

Example 1 Add $6 + (-8) + 13 + (-9)$.

Solution 1 Add the numbers in order from left to right.

$$6 + (-8) + 13 + (-9)$$
$$-2 \quad + 13 + (-9)$$
$$11 \quad\quad + (-9)$$
$$2$$

Solution 2

1. Add positive numbers.	2. Add negative numbers.	3. Add the results.
6	−8	19
13	−9	−17
19	−17	2

Add.

1.	6 2	2.	−4 −7	3.	−7 6	4.	−3 8	5.	23 64	6.	−56 31

7.	−37 −56	8.	−35 120	9.	126 −35 −37 −17	10.	−145 309 −47 −82	11.	136 −58 −47 −23	12.	−162 323 −47 −82

Add.

13. $(-8 + 5) + 2$

14. $(-12 + 15) + 6$

15. $(-4 + 8) + (-3)$

16. $(-2 + 6) + (-4)$

17. $-5 + (-3) + 5$

18. $-4 + (-14) + 4$

–3 Rules for Addition *(continued)*

Add.

19. $16 + 5 + (-8)$

20. $-6 + (-24) + 6$

21. $(-3 + 3) + 7 + (-11)$

22. $(-3 + 3) + 17 + (-7)$

23. $-2 + (-4) + (-8)$

24. $-7 + (-5) + (-6)$

25. $-3 + (-9) + 7 + (-5)$

26. $-15 + 10 + (-3) + (-2)$

Example 2 Simplify $3 + (-5) + (-x) + 7$.

Solution $3 + (-5) + (-x) + 7 = -x + \underline{3 + 7} + (-5)$ Regroup the terms.

$\quad\qquad = -x + \underbrace{\underline{10} + (-5)}$ Simplify.

$\quad\qquad = -x + \qquad 5$

Simplify.

27. $-2 + x + (-6) + 3$

28. $3 + (-8) + (-y) + (-11)$

29. $-5 + 2a + 3 + (-3)$

30. $-5 + 2a + 8 + 7$

31. $17 + 8b + (-15) + (-10)$

32. $-[6 + (-1)] + (-c) + 2$

33. $-(-7) + 3y + (-6) + 4$

34. $3x + [7 + (-2) + (-3)]$

Example 3 Evaluate $x + y + (-2)$ if $x = -2$, and $y = 5$.

Solution $x + y + (-2) = \underbrace{(-2) + 5} + (-2)$ Substitute -2 for x and 5 for y.

$\quad\qquad = \underbrace{3} + (-2)$ Add from left to right.

$\quad\qquad = \qquad 1$ Simplify.

Evaluate each expression if $x = -2$, $y = 5$, and $z = -3$.

35. $y + z + (-2)$

36. $-18 + x + y$

37. $-11 + (-x) + (-y)$

38. $-z + (-7) + y$

39. $1 + (-y) + x$

40. $-x + (-y) + (-15)$

Mixed Review Exercises

Simplify.

1. $3 + 8 \div 2$

2. $7 \cdot 5 \cdot 3 \cdot 2$

3. $(9 - 6 \div 3) \cdot 2$

4. $|-9| - 7$

5. $|-1.6| + 1.6$

6. $|-11| - |-5|$

7. $\dfrac{9 \cdot 6 + 9 \cdot 4}{6 + 3}$

8. $3\frac{1}{5} + 7\frac{1}{2} + 8\frac{4}{5}$

9. $2.7 + 1.0 + 3.3$

10. $[12 + (-2)] + 5$

11. $(-7 + 2) + (-3)$

12. $-2 + (-8) + 7 + (-1)$

2–4 Subtracting Real Numbers

Objective: To subtract real numbers and to simplify expressions involving differences.

Definition of Subtraction

To subtract a real number b, add the opposite of b.

$$a - b = a + (-b)$$

For example, $3 - 9 = 3 + (-9) = -6$.

Example 1   Simplify: **a.** $2 - 7$ **b.** $-6 - 3$ **c.** $-2 - (-8)$

Solution   **a.** $2 - 7 = 2 + (-7) = -5$

   b. $-6 - 3 = -6 + (-3) = -9$

   c. $-2 - (-8) = -2 + 8 = 6$

CAUTION 1   Subtraction is *not* commutative.

$$7 - 3 = 4,$$
$$\text{but } 3 - 7 = -4,$$
$$\text{so } 7 - 3 \neq 3 - 7$$

CAUTION 2   Subtraction is *not* associative.

$$(7 - 3) - 2 = 4 - 2 = 2,$$
$$\text{but } 7 - (3 - 2) = 7 - 1 = 6,$$
$$\text{so } (7 - 3) - 2 \neq 7 - (3 - 2)$$

Simplify.

1. $25 - 9$ **2.** $17 - 11$ **3.** $9 - 13$

4. $6 - 16$ **5.** $0 - 5$ **6.** $0 - (-3)$

7. $-12 - 0$ **8.** $-8 - 1$ **9.** $3 - (-3)$

10. $7 - (-5)$ **11.** $-8 - (-3)$ **12.** $36 - 216$

13. $143 - 270$ **14.** $36 - (-34)$ **15.** $-25 - (-24)$

16. $-15 - (-3)$ **17.** $-3 - (-15)$

18. $-17 - (-8)$ **19.** $-2.3 - 3.5$

20. $-4.2 - 5.6$ **21.** $2.65 - (-2.35)$

22. -15 decreased by 5 **23.** -8 decreased by -14

24. 18 less than -2 **25.** 10 less than -6

26. $56 - (45 - 32)$ **27.** $125 - (160 - 35)$

28. $214 - (54 - 66)$ **29.** $167 - (20 - 45)$

30. $(25 - 32) - (44 - 55)$ **31.** $(46 - 50) - (65 - 40)$

32. $(2 - 7) - (-12 + 15)$ **33.** $(32 - 24) - (-6 + 9)$

–4 Subtracting Real Numbers *(continued)*

Example 2   Simplify $13 - 9 - 8 + 5$.

Solution
$$
\begin{aligned}
13 - 9 - 8 + 5 &= 13 - 9 - 8 + 5 \\
&= \underbrace{13 + (-9)} + (-8) + 5 \\
&= \underbrace{4 \qquad + (-8)} + 5 \\
&= \underbrace{-4 \qquad\quad + 5} \\
&= 1
\end{aligned}
$$

implify.

4. $3 - 4 + 7 - 15 + 21$ **35.** $14 - 12 + 11 + 3 - 20$

6. $-5 - 18 + 6 - 7 + 10$ **37.** $-9 - 21 + 3 - 8 + 30$

Example 3   Simplify: **a.** $-(x - 5)$ **b.** $-(3 - y)$ **c.** $-(-2 + a)$

Solution   To find the opposite of a sum or a difference, you change the sign of each term of the sum or difference.

   a. $-(x - 5) = -x + 5$ **b.** $-(3 - y) = -3 + y$

   c. $-(-2 + a) = 2 - a$

implify.

8. $-(x + 2)$ **39.** $-(4 - y)$ **40.** $-(-7 + a)$ **41.** $-(x - 3)$

2. $-(y - 5)$ **43.** $-(8 - x)$ **44.** $-(b - 6)$ **45.** $-(2 + n)$

Example 4   Simplify $8 - (x + 3)$.

Solution
$$
\begin{aligned}
8 - (x + 3) &= 8 - x - 3 \qquad && \text{Change the sign of each term of } x + 3. \\
&= (8 - 3) - x \qquad && \text{Regroup the terms.} \\
&= 5 - x && \text{Simplify.}
\end{aligned}
$$

implify.

6. $6 - (y + 4)$ **47.** $4 - (q - 6)$ **48.** $x - (x + 2)$ **49.** $n - (-3 + n)$

ixed Review Exercises

1. $|-6| + |2|$ **2.** $17 \cdot 2 \cdot 3 \cdot 5$ **3.** $2 + 6x + 5y + 8$

4. $\left|-\dfrac{3}{4}\right| - \left|-\dfrac{1}{4}\right|$ **5.** $-\dfrac{3}{2} + \left(-\dfrac{5}{2}\right)$ **6.** $1\dfrac{1}{4} + \left(-3\dfrac{3}{4}\right)$

7. $[5 + (-9)] + 7$ **8.** $3.4 - 0.5 + (-1.4)$ **9.** $-4 + [-6 + (-2)]$

0. $-2.4 + 8.3 + (-3.6)$ **11.** $-27 + (-28) + 18 + 47$ **12.** $2 + (-3) + (-10) + (-x)$

NAME _____ DATE _____

2–5 The Distributive Property

Objective: To use the distributive property to simplify expressions.

Vocabulary

Equivalent expressions Expressions that represent the same number.

Simplifying an expression Replacing an expression containing variables by
   an equivalent expression with as few terms as possible.

Distributive Property

Distributive Property of Multiplication (with respect to addition)
For all real numbers a, b, and c, $a(b + c) = ab + ac$ and $(b + c)a = ba + ca$.
For example, $6(9 + 4) = 6 \cdot 9 + 6 \cdot 4$ and $(9 + 4)6 = 9 \cdot 6 + 4$

Distributive Property of Multiplication (with respect to subtraction)
For all real numbers a, b, and c, $a(b - c) = ab - ac$ and $(b - c)a = ba - ca$.
For example, $8(12 - 2) = 8 \cdot 12 - 8 \cdot 2$ and $(12 - 2)8 = 12 \cdot 8 - $

CAUTION   When using the distributive properties, be sure to multiply
   both of the numbers inside the parentheses by the number
   outside the parentheses. For example,
   $6(13 - 3) = 6 \cdot 13 - 6 \cdot 3$ *not* $6 \cdot 13 - 3$.

Example 1   Simplify: **a.** $5 \cdot 48$ **b.** $8(7.5)$ **c.** $6\left(4\frac{1}{3}\right)$ **d.** $(11 - 5)9$

Solution   Use the distributive property to multiply.

a. $5 \cdot 48 = 5(40 + 8)$ **b.** $8(7.5) = 8(7 + 0.5)$
   $= (5 \cdot 40) + (5 \cdot 8)$ $= (8 \cdot 7) + (8 \cdot 0.5)$
   $= 200 + 40$ $= 56 + 4$
   $= 240$ $= 60$

c. $6\left(4\frac{1}{3}\right) = 6\left(4 + \frac{1}{3}\right)$ **d.** $(11 - 5)9 = (11 - 5)9$
   $= (6 \cdot 4) + \left(6 \cdot \frac{1}{3}\right)$ $= (11 \cdot 9) - (5 \cdot 9)$
   $= 24 + 2$ $= 99 - 45$
   $= 26$ $= 54$

Simplify. Use the distributive property.

1. $6 \cdot 35$ **2.** $5 \cdot 52$ **3.** $4(8.5)$ **4.** $8(6.25)$

5. $12\left(2\frac{1}{3}\right)$ **6.** $10\left(2\frac{1}{5}\right)$ **7.** $15\left(3\frac{2}{3}\right)$ **8.** $12 \cdot 25$

9. $5(20 - 1)$ **10.** $6(60 - 2)$ **11.** $9(30 - 1)$ **12.** $8(40 - 3)$

13. $(9 - 4)6$ **14.** $(12 - 3)8$ **15.** $(20 - 1)5$ **16.** $(30 - 7)6$

---

5 *The Distributive Property* (continued)

Example 2   Simplify: **a.** $64 \cdot 19 + 36 \cdot 19$ **b.** $(3.8)(25) - (1.8)(25)$

Solution   **a.** $64 \cdot 19 + 36 \cdot 19 = (64 + 36)19$ **b.** $(3.8)(25) - (1.8)(25) = (3.8 - 1.8)25$
$\qquad\qquad\qquad\qquad = (100)19 \qquad\qquad\qquad\qquad\qquad\qquad = (2)25$
$\qquad\qquad\qquad\qquad = 1900 \qquad\qquad\qquad\qquad\qquad\qquad\ \ = 50$

plify.

20. $19 + 80 \cdot 19$ **18.** $13 \cdot 53 + 87 \cdot 53$ **19.** $(17 \cdot 24) - (17 \cdot 24)$

    $(63 \cdot 71) + (37 \cdot 71)$ **21.** $(0.65)(28) + (0.35)(28)$ **22.** $(4.3)(25) - (2.3)(25)$

Example 3   Write an equivalent expression without parentheses.

   a. $5(n - 2)$ **b.** $(5y + 6)2$

Solution   **a.** $5(n - 2) = 5 \cdot n - 5 \cdot 2$ **b.** $(5y + 6)2 = (5y)2 + (6)2$
$\qquad\qquad\qquad\qquad = 5n - 10 \qquad\qquad\qquad\qquad\quad\ = 10y + 12$

each expression write an equivalent expression without parentheses.

 $2(x + 3)$ **24.** $6(a + 5)$ **25.** $5(n - 1)$ **26.** $7(b - 5)$

 $3(6n + 2)$ **28.** $8(5n - 3)$ **29.** $3(x - y)$ **30.** $2(4x - y)$

 $(4n - 7)4$ **32.** $(3x + 4)5$ **33.** $(3x + 4y)8$ **34.** $(5m + 7n)2$

Example 4   Simplify: **a.** $8x + 6x$ **b.** $9y - 2y$ **c.** $5n - 6 + 3n$

Solution   **a.** $8x + 6x = (8 + 6)x = 14x$
   b. $9y - 2y = 9y + (-2)y = (9 - 2)y = 7y$
   c. $5n - 6 + 3n = 5n + 3n - 6 = 8n - 6$

mplify.

 $6a + 4a$ **36.** $5m + 7m$ **37.** $15y - 6y$ **38.** $3x - (-9)x$

 $(-4)n + 9n$ **40.** $(-7)n - 8n$ **41.** $2a + 9 + 5a$ **42.** $7n + 1 + 3n$

 $9n - 5 + 2n$ **44.** $3x + 8 - 2x$ **45.** $9y - 6 + 5y$ **46.** $10n - 7 + 6n$

ixed Review Exercises

aluate if $a = -2$, $b = -3$, $c = 4$, $x = 6$, and $y = 8$.

 $4x + y - c$ **2.** $(x \cdot x + c) \div 8$ **3.** $3y - (2x \div c)$ **4.** $|a| + |b| + (-y)$

 $c + |a| + |-y|$ **6.** $2|a| - 3|b|$ **7.** $-(x - b) + c$ **8.** $x + y + (-5)$

 $a + b + (-c)$ **10.** $c - (a - b)$ **11.** $-a + b - c$ **12.** $|b - a| - c$

2–6 *Rules for Multiplication*

Objective: To multiply real numbers.

Properties	Examples
Identity Property of Multiplication The product of a number and 1 is identical to the number itself. $\quad a \cdot 1 = a \quad$ and $\quad 1 \cdot a = a$	$6 \cdot 1 = 6$ and $1 \cdot 6 = 6$
Multiplication Property of Zero When one of the factors of a product is zero, the product itself is zero. $\quad a \cdot 0 = 0 \quad$ and $\quad 0 \cdot a = 0$	$6 \cdot 0 = 0$ and $0 \cdot 6 = 0$
Multiplication Property of -1 For every real number a: $\quad a(-1) = -a \quad$ and $\quad (-1)a = -a$	$6(-1) = -6$ and $(-1)6 = -6$ $(-5)(-1) = -(-5) = 5$ and $(-1)(-5) = -(-5) = 5$
Property of Opposites in Products For all real numbers a and b: $\quad (-a)(b) = -ab$ $\quad a(-b) = -ab$ $\quad (-a)(-b) = ab$	$(-4)(5) = -20$ $4(-5) = -20$ $(-4)(-5) = 20$

Rules for Multiplication

1. If two numbers have the *same* sign, their product is positive.
   If two numbers have *opposite* signs, their product is negative.

2. The product of an *even* number of negative numbers is positive.
   The product of an *odd* number of negative numbers is negative.

Example 1　　Multiply:　　**a.** 3(6)　　**b.** $(-3)(6)$　　**c.** 3(-6)　　**d.** $(-3)(-6)$

Solution　　**a.** $3(6) = 18$　　　　(Both factors have the same sign.)

　　　　　　　b. $(-3)(6) = -18$　　(The two factors have opposite signs.)

　　　　　　　c. $3(-6) = -18$　　　(The two factors have opposite signs.)

　　　　　　　d. $(-3)(-6) = 18$　　(Both factors have the same sign.)

Example 2　　**a.** $2(-3)(-4)(-5)$ is negative because it has 3 negative factors.

　　　　　　　b. $(-1)(-4)(-5)(6)(-7)$ is positive because it has 4 negative factors.

　　　　　　　c. $(-6)(7)(0)(-4)$ is zero because it has a zero factor.

-6 Rules for Multiplication (continued)

Multiply.

1. $(-12)(-3)$ 2. $18(-4)$ 3. $2(17)$ 4. $18(0)$

5. $(-2)(5)(-8)$ 6. $(4)(-7)(10)$ 7. $(-2)(-3)(-4)$ 8. $(-11)(-12)(0)$

9. $35(-26)(0)$ 10. $5(-2)(-8)(-5)$ 11. $(-7)(3)(-1)(2)$ 12. $(-8)(-5)(-1)(-3)$

Example 3 Simplify: **a.** $(-2x)(-6y)$ **b.** $3y + (-7y)$

Solution **a.** $(-2x)(-6y) = (-2)x(-6)y$ **b.** $3y + (-7y) = [3 + (-7)]y$
$\qquad\qquad\quad\ = (-2)(-6)xy \qquad\qquad\qquad\qquad = (-4)y$
$\qquad\qquad\quad\ = 12xy \qquad\qquad\qquad\qquad\qquad\ = -4y$

Simplify.

13. $(-3a)(-4b)$ 14. $(5x)(6y)$ 15. $2p(-5q)$ 16. $(-4e)(7f)$ 17. $(-6a)(-5b)$

18. $-7a + (-8a)$ 19. $2x + (-5x)$ 20. $8x + (-3x)$ 21. $(-11y) + 3y$ 22. $-4n + 4n$

Example 4 Simplify: **a.** $-3(2x - y)$ **b.** $5x - 4(x - 1)$

Solution **a.** $-3(2x - y) = -3(2x) - (-3)(y)$ **b.** $5x - 4(x - 1) = 5x - (4x - 4 \cdot 1)$
$\qquad\qquad\qquad\ \ = -6x - (-3y) \qquad\qquad\qquad\qquad\ \ = 5x - (4x - 4)$
$\qquad\qquad\qquad\ \ = -6x + 3y \qquad\qquad\qquad\qquad\qquad = 5x - 4x + 4$
$\qquad\qquad\qquad\qquad\qquad\qquad\qquad\qquad\qquad\qquad\quad\ = x + 4$

Simplify.

23. $-6(x - 2y)$ 24. $-5(2c + d)$ 25. $-4(3m + 2n)$

26. $-7(-4y - 5)$ 27. $(3x - 5)(-6)$ 28. $(-3 + 5y)(-2)$

29. $4x - 3(x - 2)$ 30. $6x - 2(x + 3)$ 31. $3x - 5(x - 1)$

32. $(-1)(a - b + 2)$ 33. $(-1)(2x - y - 3)$ 34. $(-1)(x + y - z)$

35. $4x - 2x + 7 + x$ 36. $2y - 5 - 5y + 3$ 37. $11p - 6c - 7c + 9p$

Mixed Review Exercises

Translate each sentence into an equation.

1. Three times a number is 27.   2. The quotient of n and 4 is 15.

3. One half of a number is nine.   4. Six less than twice a number is 14.

Simplify.

5. $110 - (12 - 8)$ 6. $161 - (8 - 11)$ 7. $2 + (-5) + (-y) + 9$

8. $3(20 + 5)$ 9. $2n + (-5n)$ 10. $5(n + 1) + 7$

2-7 Problem Solving: Consecutive Integers

Objective: To write equations to represent relationships among integers.

Vocabulary

Consecutive Integers Numbers obtained by counting by ones from any integer.
For example, $-2, -1, 0, 1,$ and 2 are consecutive integers.

Even Integer An integer that is the product of 2 and any integer.
For example, $-10, -4, 2, 6,$ and 8 are even integers.

Odd Integer An integer that is not even.
For example, $-3, -1, 7, 9,$ and 11 are odd integers.

Consecutive Even Integers Numbers obtained by counting by twos from any even integer.
For example, $-6, -4, -2, 0,$ and 2 are consecutive even integers.

Consecutive Odd Integers Numbers obtained by counting by twos from any odd integer.
For example, $-5, -3, -1, 1,$ and 3 are consecutive odd integers.

Example 1   An integer is represented by n.
  a. Write the next four consecutive integers after n.
  b. Write an equation to represent this relationship:
      The sum of three consecutive integers starting with n is 93.
  c. Write an equation to represent this relationship:
      The product of two consecutive integers starting with n is 56.

Solution   **a.** $n + 1, n + 2, n + 3, n + 4$
  b. $n + (n + 1) + (n + 2) = 93$
  c. $n(n + 1) = 56$

Write an equation to represent the given relationship.

1. The sum of three consecutive integers is 39.

2. The sum of three consecutive integers is -51.

3. The product of two consecutive integers is 42.

4. The product of two consecutive integers is 30.

Example 2   Write an equation to represent this relationship:
The sum of three consecutive odd numbers is 33.

Solution   Let $n =$ the first integer, $n + 2 =$ the second integer,
and $n + 4 =$ the third integer.

$$\underbrace{\text{The sum of three consecutive odd integers}}_{n + (n + 2) + (n + 4)} \underset{=}{\overset{\text{is 33.}}{\downarrow}} 33$$

Write an equation to represent the given relationship.

5. The sum of three consecutive odd integers is 45.

6. The sum of three consecutive even integers is

7. The sum of four consecutive integers is 90.

8. The sum of four consecutive even integers is

-7 *Problem solving: Consecutive Integers* (continued)

Example 3	Write an equation to represent this relationship: The product of two consecutive integers is 110.
Solution	Let n = the first integer and $n + 1$ = the second integer.

The product of the two consecutive integers is 110.

$$n(n + 1) \qquad = 110$$

Write an equation to represent the given relationship.

9. The product of two consecutive integers is 72.

10. The product of two consecutive even integers is 80.

11. The product of two consecutive integers is 132.

12. The product of two consecutive odd integers is 195.

Example 4		Solve over the given domain: The sum of three consecutive odd integers is 32 more than the smallest integer. What are the integers? Domain for the smallest integer: {9, 11, 13}
Solution	*Step 1*	The unknowns are the three consecutive odd integers.
	Step 2	Let n = the smallest integer, $n + 2$ = the middle integer, and $n + 4$ = the largest integer.

Step 3

The sum is 32 more than the smallest integer.

$$n + (n + 2) + (n + 4) = n + 32$$

Step 4   Replace n in turn by 9, 11, and 13.

n	$n + (n + 2) + (n + 4)$	$=$	$n + 32$	
9	$9 + 11 + 13$	$=$	$9 + 32$	False
11	$11 + 13 + 15$	$=$	$11 + 32$	False
13	$13 + 15 + 17$	$=$	$13 + 32$	True

Step 5   The check is left to you. The integers are 13, 15, and 17.

Solve over the given domain.

13. The sum of three consecutive even integers is 50 more than the largest integer. What are the integers? Domain for the smallest integer: {20, 22, 24}

14. The sum of three consecutive odd integers is 72 more than the smallest integer. What are the integers? Domain for the smallest integer: {29, 31, 33}

Mixed Review Exercises

1. $(30 - 3) - (43 - 20)$

2. $-4.5 + 2.3 - 1.7$

3. $-2 + 3c + (-4) + 7$

4. $\dfrac{7}{4} + \left(-\dfrac{10}{4}\right)$

5. $2\dfrac{2}{5} + 10 + 3\dfrac{3}{5}$

6. $5\left(\dfrac{3}{4}\right) - 4\left(\dfrac{1}{4}\right) + 3\left(\dfrac{3}{4}\right)$

7. $-(10 - x) - (x - 15)$

8. $14a - 4a + 5a$

9. $10 + 4y + 5 + (-3)$

2–8 *The Reciprocal of a Real Number*

Objective: To simplify expressions involving reciprocals.

Vocabulary

Reciprocals Two numbers whose product is 1 are called reciprocals of each other.
For example, 5 and $\frac{1}{5}$ are reciprocals.

Symbols $\frac{1}{a}$ (the reciprocal of a) $-\frac{1}{a}$ (the reciprocal of $-a$)

Properties	Examples
Property of Reciprocals Every *nonzero* real number a has a reciprocal $\frac{1}{a}$, such that $\qquad a \cdot \frac{1}{a} = 1 \quad$ and $\quad \frac{1}{a} \cdot a = 1.$	$3 \cdot \frac{1}{3} = 1 \quad$ and $\quad \frac{1}{3} \cdot 3 = 1$
Property of the Reciprocal of the Opposite of a Number For every *nonzero* number a, $\qquad \frac{1}{-a} = -\frac{1}{a}.$	$\frac{1}{-3} = -\frac{1}{3}$
Property of the Reciprocal of a Product For all *nonzero* numbers a and b, $\qquad \frac{1}{ab} = \frac{1}{a} \cdot \frac{1}{b}.$	$\frac{1}{2 \cdot 3} = \frac{1}{2} \cdot \frac{1}{3}$

CAUTION   0 has no reciprocal; 1 is its own reciprocal; and -1 is its own reciprocal.

Example 1   Simplify: **a.** $\frac{1}{3} \cdot \frac{1}{-5}$ **b.** $3y \cdot \frac{1}{3}$ **c.** $(-6xy)\left(-\frac{1}{2}\right)$

Solution   **a.** $\frac{1}{3} \cdot \frac{1}{-5} = \frac{1}{3(-5)} = \frac{1}{-15} = -\frac{1}{15}$

   b. $3y \cdot \frac{1}{3} = \left(3 \cdot \frac{1}{3}\right)y = 1y = y$

   c. $(-6xy)\left(-\frac{1}{2}\right) = (-6)\left(-\frac{1}{2}\right)(xy) = 3xy$

Simplify each expression.

1. $\frac{1}{3}(-12)$

2. $-\frac{1}{8}(24)$

3. $-50\left(\frac{1}{5}\right)$

4. $-30\left(\frac{1}{3}\right)$

5. $(-20)\left(-\frac{1}{4}\right)$

6. $(-42)\left(-\frac{1}{7}\right)$

7. $-36\left(-\frac{1}{4}\right)\left(\frac{1}{3}\right)$

8. $60\left(-\frac{1}{5}\right)\left(-\frac{1}{12}\right)$

9. $72\left(-\frac{1}{8}\right)\left(-\frac{1}{9}\right)$

10. $-54\left(-\frac{1}{6}\right)\left(-\frac{1}{9}\right)$

11. $\frac{1}{-2}(24)\left(\frac{1}{4}\right)$

12. $-60\left(\frac{1}{2}\right)\left(\frac{1}{3}\right)$

–8 The Reciprocal of a Real Number *(continued)*

Simplify each expression.

3. $6r\left(-\dfrac{1}{6}\right)$ **14.** $32p\left(-\dfrac{1}{8}\right)$ **15.** $\dfrac{1}{x}(8x),\ x \neq 0$ **16.** $9x\left(\dfrac{1}{x}\right),\ x \neq 0$

7. $21xy\left(\dfrac{1}{7}\right)$ **18.** $72ab\left(\dfrac{1}{9}\right)$ **19.** $18xy\left(\dfrac{1}{6}\right)$ **20.** $(-54xy)\left(\dfrac{1}{-9}\right)$

1. $15xy\left(\dfrac{1}{-3}\right)$ **22.** $6cd\left(\dfrac{1}{-2}\right)$ **23.** $(-8pq)\left(\dfrac{1}{-2}\right)$ **24.** $(-42ac)\left(\dfrac{1}{-7}\right)$

Example 2   Simplify: **a.** $\dfrac{1}{2}(8m - 4n)$ **b.** $(-21a - 63b)\left(-\dfrac{1}{7}\right)$

Solution   **a.** $\dfrac{1}{2}(8m - 4n) = \dfrac{1}{2}(8m) - \dfrac{1}{2}(4n)$ Use the distributive property.

$= \left(\dfrac{1}{2}\cdot 8\right)m - \left(\dfrac{1}{2}\cdot 4\right)n$ Use the associative property.

$= 4m - 2n$ Simplify.

b. $(-21a - 63b)\left(-\dfrac{1}{7}\right) = (-21a)\left(-\dfrac{1}{7}\right) - (63b)\left(-\dfrac{1}{7}\right)$

$= (-21)\left(-\dfrac{1}{7}\right)a - (63)\left(-\dfrac{1}{7}\right)b$

$= (3)a - (-9)b$

$= 3a + 9b$

Simplify each expression.

5. $\dfrac{1}{2}(-8a + 10)$ **26.** $\dfrac{1}{3}(9y - 21)$

. $-\dfrac{1}{5}(-25c + 10d)$ **28.** $-\dfrac{1}{4}(24g - 32h)$

. $(-21m - 14n)\left(-\dfrac{1}{7}\right)$ **30.** $(-26e - 52f)\left(-\dfrac{1}{13}\right)$

. $(40x - 56y)\left(-\dfrac{1}{8}\right)$ **32.** $(-5a + 30b)\left(\dfrac{1}{-5}\right)$

Mixed Review Exercises

Translate each sentence into an equation.

1. Three more than six times a number is 21.   **2.** Twelve less than a number is 200.

3. The sum of two consecutive integers is 71.   **4.** The product of two consecutive integers is 90.

Simplify.

5. $(-8)(-3)(-5)$ **6.** $-24(25)(-4)$ **7.** $5(-7)(-6)$

8. $-8(2a - 5b)$ **9.** $-2(2 + x) - 2(x - 2)$ **10.** $10(x - 1) + 4(3 - x)$

2-9 Dividing Real Numbers

Objective: To divide real numbers and to simplify expressions involving quotients.

Definition of Division

To divide by a nonzero real number b, multiply by the reciprocal of b.

$$a \div b \text{ or } \frac{a}{b} = a \cdot \frac{1}{b}. \qquad \text{For example, } 24 \div 3 = 24 \cdot \frac{1}{3}.$$

Rules for Division

If two numbers have the same sign, their quotient is positive.
If two numbers have opposite signs, their quotient is negative.

CAUTION 1 You can't divide by zero since zero has no reciprocal.

CAUTION 2 Division is not commutative. For example, $4 \div 2 = 2$, but $2 \div 4 = \frac{1}{2}$.

CAUTION 3 Division is not associative. For example, $(12 \div 6) \div 2 = 2 \div 2 = 1$,
but $12 \div (6 \div 2) = 12 \div 3 = 4$.

Example 1 Simplify: **a.** $\frac{30}{6}$ **b.** $\frac{30}{-6}$ **c.** $\frac{-30}{6}$ **d.** $\frac{-30}{-6}$

Solution   **a.** $\frac{30}{6} = 30 \div 6 = 30 \cdot \frac{1}{6} = 5$ **b.** $\frac{30}{-6} = 30 \div (-6) = 30\left(-\frac{1}{6}\right) = -5$

c. $\frac{-30}{6} = -30 \div 6 = -30 \cdot \frac{1}{6} = -5$ **d.** $\frac{-30}{-6} = -30 \div (-6) = -30\left(-\frac{1}{6}\right) = 5$

Simplify.

1. $42 \div 14$ **2.** $-56 \div 7$ **3.** $-24 \div (-6)$ **4.** $-32 \div (-8)$

5. $\frac{-144}{12}$ **6.** $\frac{96}{-16}$ **7.** $\frac{-100}{-5}$ **8.** $\frac{-75}{-3}$

Example 2 Simplify: **a.** $8 \div \left(-\frac{4}{5}\right)$ **b.** $\frac{-4}{-\frac{1}{2}}$

Solution   **a.** $8 \div \left(-\frac{4}{5}\right) = 8\left(-\frac{5}{4}\right) = -10$ **b.** $\frac{-4}{-\frac{1}{2}} = (-4) \div \left(-\frac{1}{2}\right) = (-4)(-2) = 8$

Simplify.

9. $6 \div \left(-\frac{1}{3}\right)$ **10.** $12 \div \left(-\frac{1}{4}\right)$ **11.** $0 \div \frac{5}{6}$ **12.** $-8 \div \left(-\frac{1}{2}\right)$ **13.** $0 \div \left(-\frac{2}{7}\right)$

14. $\frac{-12}{-\frac{1}{4}}$ **15.** $\frac{8}{-\frac{1}{2}}$ **16.** $\frac{-20}{\frac{1}{5}}$ **17.** $\frac{0}{\frac{1}{9}}$ **18.** $\frac{-8}{-\frac{1}{8}}$

NAME _____   DATE _____

2–9 Dividing Real Numbers (continued)

Example 3 Simplify: **a.** $\dfrac{32x}{-8}$ **b.** $\dfrac{w}{12} \cdot 12$

Solution   **a.** $\dfrac{32x}{-8} = 32x\left(-\dfrac{1}{8}\right)$ Multiply by the reciprocal of -8.

$\phantom{a.\ \dfrac{32x}{-8}} = 32\left(-\dfrac{1}{8}\right)x$ Regroup the factors.

$\phantom{a.\ \dfrac{32x}{-8}} = -4x$ Simplify.

b. $\dfrac{w}{12} \cdot 12 = w \cdot \dfrac{1}{12} \cdot 12$

$\phantom{b.\ \dfrac{w}{12} \cdot 12} = w \cdot 1$

$\phantom{b.\ \dfrac{w}{12} \cdot 12} = w$

Simplify.

19. $\dfrac{-18x}{3}$

20. $\dfrac{-42x}{6}$

21. $\dfrac{50x}{-10}$

22. $\dfrac{-36x}{-6}$

23. $5 \cdot \dfrac{x}{5}$

24. $\dfrac{-w}{8} \cdot 8$

25. $(-6)\left(\dfrac{-y}{2}\right)$

26. $(-10)\left(\dfrac{x}{-2}\right)$

27. $\dfrac{144b}{12}$

28. $\dfrac{121b}{-11}$

29. $\dfrac{-48x}{6}$

30. $\dfrac{-108x}{-36}$

Example 4 Find the average of $14, -2, -8, -12$.

Solution Find the sum of the numbers and divide by the number of numbers.

$$\dfrac{14 + (-2) + (-8) + (-12)}{4} = \dfrac{-8}{4} = -2$$

Find the average of the given numbers.

31. $-12, 5, -10, -7$ **32.** $15, -21, -8, 6$ **33.** $13, -5, -16, -4$

34. $23, -13, -18, 20$ **35.** $11, -15, -22, 16, 0$ **36.** $23, -12, -17, 21, 5$

Mixed Review Exercises

Solve if $x \in \{0, 1, 2, 3, 4, 5, 6\}$.

1. $x + 5 = 7$ **2.** $\dfrac{1}{2}x = 3$ **3.** $x - 1 = 4$

4. $3x = 9$ **5.** $3x + 1 = 7$ **6.** $x \div 3 = 1$

Solve over the domain $\{0, 1, 2, 3, 4, 5\}$.

7. $\dfrac{1}{3}n = 1$ **8.** $3y - 1 = 14$ **9.** $x + 2 = 6$

10. $2x = 2$ **11.** $x \cdot x = 4$ **12.** $3n = n \cdot 3$

3 Solving Equations and Problems

3–1 *Transforming Equations: Addition and Subtraction*

Objective: To solve equations using addition and subtraction.

Properties

Addition Property of Equality If the same number is added to equal numbers, the sums are equal.

Subtraction Property of Equality If the same number is subtracted from equal numbers, the differences are equal.

Vocabulary

Equivalent equations Equations that have the same solution set over a given domain.

Transformations Operations on an equation that produce a simpler equivalent equation.

By substitution You can substitute an equivalent expression for any expression in an equation. You do this when you simplify an expression in an equation.

By addition You can add the same number to each side of an equation.

By subtraction You can subtract the same number from each side of an equation.

CAUTION

To check your work, you should check that each solution of the final equation satisfies the *original equation.*

Example 1 Solve $x - 6 = 11$.

Solution

$$x - 6 = 11$$
$$x - 6 + 6 = 11 + 6$$
$$x = 17$$

{ To get x alone on one side, add 6 to each side and then simplify.

Check: $x - 6 = 11$ ——— Original equation.

$17 - 6 \overset{?}{=} 11$ Substitute 17 for x.

$11 = 11$ ✓

The solution set is $\{17\}$.

Solve.

1. $a - 9 = 11$

2. $b - 5 = 13$

3. $x - 20 = -19$

4. $d - 14 = 5$

5. $x - 15 = 0$

6. $v - 27 = -54$

7. $x - 6 = 27$

8. $q - 7 = 11$

9. $q - 9 = -16$

-1 Transforming Equations: Addition and Subtraction *(continued)*

Example 2   Solve $-9 = n + 11$.

Solution

$$-9 = n + 11$$
$$-9 - 11 = n + 11 - 11$$
$$-20 = n$$

{ To get n alone on one side,
subtract 11 from each side.

Simplify.

Check: $-9 = n + 11$ ◄——— Original equation

$-9 \overset{?}{=} -20 + 11$ Substitute -20 for n.

$-9 = -9 \;\surd$ The solution set is $\{-20\}$.

Solve.

. $-6 = m + 6$	**11.** $21 = x + 15$	**12.** $-26 + m = 24$	**13.** $-37 + n = 63$
. $p + 18 = -22$	**15.** $a + 60 = -15$	**16.** $14 + t = 0$	**17.** $29 = y - 12$
. $35 = x + 16$	**19.** $-4 = u - 6$	**20.** $22 = y + 3$	**21.** $c - 8 = -10$
. $x + 1.5 = 6.8$	**23.** $-1 + a = 0.5$	**24.** $3.9 = y - 1.4$	**25.** $7.5 = w - 2.5$

Example 3   Solve $-x + 5 = 4$.

Solution

$$-x + 5 = 4$$
$$-x + 5 - 5 = 4 - 5$$
$$-x = -1$$
$$x = 1$$

{ To get x alone on one side,
subtract 5 from each side and simplify.

{ If the opposite of a number is -1,
the number must be 1.

Check: $-x + 5 = 4$ ◄——— Original equation

$-1 + 5 \overset{?}{=} 4$ Substitute 1 for x.

$4 = 4 \;\surd$ The solution set is $\{1\}$.

Solve.

. $-x + 3 = 5$	**27.** $-y + 7 = 17$	**28.** $12 - x = 18$
. $7 - y = 11$	**30.** $9 = -x + 16$	**31.** $13 = 22 - y$
. $-5 - y = 7$	**33.** $10 = -12 - e$	**34.** $15 = -y + 10$

Mixed Review Exercises

Evaluate if $a = 3$, $b = -6$, $c = -4$, and $d = 2$.

. $a - \lvert b - c \rvert$	**2.** $(\lvert c \rvert - d) - (\lvert b \rvert - a)$	**3.** $3\lvert c \rvert - (-b)$
. $\dfrac{a - 2b}{a + d}$	**5.** $\dfrac{3b + c - d}{ad}$	**6.** $\dfrac{2ab}{c + d}$

Simplify.

. $(-3)(-4)(8)$	**8.** $(-7 \cdot 16) + (-7 \cdot 24)$	**9.** $252 \div (-36)$

3–2 *Transforming Equations: Multiplication and Division*

Objective: To solve equations using multiplication or division.

Properties

Multiplication Property of Equality If equal numbers are multiplied by the same number, the products are equal.

Division Property of Equality If equal numbers are divided by the same *nonzero* number, the quotients are equal.

Transformations

By multiplication You can multiply each side of an equation by the same *nonzero* real number.

By division You can divide each side of an equation by the same *nonzero* real number.

CAUTION 1 When you transform an equation, never multiply or divide by zero.

CAUTION 2 When you multiply or divide by a negative number, be careful with the sign of your answer.

Example 1   Solve $4x = 128$.

Solution $\dfrac{4x}{4} = \dfrac{128}{4}$ $\left\{\begin{array}{l}\text{To get } x \text{ alone on one side, divide each side}\\ \text{by 4 (or multiply by } \frac{1}{4}\text{, the reciprocal of 4).}\end{array}\right.$

          $x = 32$

        Check: $4x = 128$

                $4(32) \overset{?}{=} 128$

                $128 = 128 \;\checkmark$ The solution set is $\{32\}$.

Solve.

1. $7m = 140$ **2.** $12n = 240$ **3.** $-8x = 96$

4. $-11f = -143$ **5.** $-720 = -24g$ **6.** $330 = -15u$

7. $108 = -9x$ **8.** $45k = -270$ **9.** $26n = -520$

Example 2   Solve $12 = -\dfrac{3}{4}n$.

Solution $-\dfrac{4}{3}(12) = -\dfrac{4}{3}\left(-\dfrac{3}{4}n\right)$ $\left\{\begin{array}{l}\text{To get } n \text{ alone on one side, multiply each}\\ \text{side by } -\frac{4}{3}\text{, the reciprocal of } -\frac{3}{4}.\end{array}\right.$

            $-16 = n$

        Check: $12 = -\dfrac{3}{4}n$

              $12 \overset{?}{=} -\dfrac{3}{4}(-16)$

              $12 = 12 \;\checkmark$ The solution set is $\{-16\}$.

–2 Transforming Equations: Multiplication and Division *(continued)*

Solve.

$\frac{2}{3}m = 6$

11. $\frac{3}{5}y = -15$

12. $-\frac{5}{8}x = 40$

. $-\frac{4}{5}y = -20$

14. $\frac{2}{5}d = -40$

15. $\frac{3}{4}g = -24$

. $\frac{7}{8}y = -56$

17. $-\frac{7}{10}e = 140$

18. $-\frac{2}{7}n = -28$

Example 3 Solve: **a.** $\frac{x}{2} = -6$ **b.** $\frac{1}{2}n = 3\frac{1}{2}$

Solution

$2\left(\frac{x}{2}\right) = 2(-6)$

$x = -12$

Check: $\frac{x}{2} = -6$

$\frac{-12}{2} \overset{?}{=} -6$

$-6 = -6 \checkmark$

The solution set is $\{-12\}$.

$\frac{1}{2}n = \frac{7}{2}$

$2\left(\frac{1}{2}\right)n = 2\left(\frac{7}{2}\right)$

$n = 7$

Check: $\frac{1}{2}n = 3\frac{1}{2}$

$\frac{1}{2}(7) \overset{?}{=} \frac{7}{2}$

$\frac{7}{2} = \frac{7}{2} \checkmark$

The solution set is $\{7\}$.

---

lve.

. $\frac{c}{6} = -24$

20. $\frac{y}{5} = -25$

21. $-\frac{u}{12} = 12$

. $-\frac{x}{3} = 15$

23. $-28 = \frac{c}{7}$

24. $-\frac{1}{4}x = 2\frac{1}{4}$

. $\frac{1}{5}f = 3\frac{1}{5}$

26. $\frac{1}{2}b = 2\frac{1}{2}$

27. $-\frac{1}{3}y = 3\frac{2}{3}$

ixed Review Exercises

aluate if $a = -2$, $b = 3$, and $c = -6$.

. $6b - 2a$

2. $(2b - 5c)a$

3. $|c| + |a| - b$

. $|b| - |a + c|$

5. $\frac{-(7ab)}{c}$

6. $\frac{8 + a}{c}$

nplify.

. $6a + 5 + 7a$

8. $7n - 6 + 6$

9. $9p - p + 3$

. $-3(m + 4)$

11. $(x + 5)6$

12. $2(3y - 4)$

3–3 *Using Several Transformations*

Objective: To solve equations by using more than one transformation.

Vocabulary

Inverse operations Operations that "undo" each other. For example, multiplication and division are inverse operations. Likewise, addition and subtraction are inverse operations.

Tips for solving an equation in which the variable is on one side.

1. Simplify each side of the equation as needed.
2. Use inverse operations to "undo" the operations in the equation.

Example 1   Solve $3n - 7 = 8$.

Use inverse operations:

Solution $3n - 7 + 7 = 8 + 7$ To undo the subtraction of 7 from $3n$, add 7 to each side.

$$3n = 15$$

$$\frac{3n}{3} = \frac{15}{3}$$ To undo the multiplication of n by 3, divide each side by 3

$$n = 5$$ The solution set is $\{5\}$.

Example 2   Solve $\frac{1}{2}x + 1 = 7$.

Use inverse operations:

Solution $\frac{1}{2}x + 1 - 1 = 7 - 1$ Subtract 1 from each side.

$$\frac{1}{2}x = 6$$

$$2\left(\frac{1}{2}x\right) = 6 \cdot 2$$ Multiply each side by 2, the reciprocal of $\frac{1}{2}$.

$$x = 12$$ The solution set is $\{12\}$.

Solve.

1. $2y + 1 = 15$ **2.** $2x - 7 = 13$ **3.** $26 = 5y + 1$ **4.** $58 = 3y - 2$

5. $-11 + 4y = 25$ **6.** $13 + 6y = -23$ **7.** $\frac{1}{2}x - 3 = 5$ **8.** $\frac{1}{3}x + 5 = 7$

9. $3 = \frac{1}{4}x - 1$ **10.** $6 = \frac{1}{5}x + 2$ **11.** $\frac{x}{2} + 7 = 1$ **12.** $\frac{x}{5} - 2 = 4$

Example 3   Solve $\frac{x - 2}{3} = 4$.

Solution $3\left(\frac{x - 2}{3}\right) = 3 \cdot 4$ Multiply each side by 3.

$$x - 2 = 12$$

$$x - 2 + 2 = 12 + 2$$ Add 2 to each side.

$$x = 14$$ The solution set is $\{14\}$.

–3 Using Several Transformations (continued)

Solve.

13. $\dfrac{x-1}{2} = 5$

14. $\dfrac{3-x}{4} = 2$

15. $\dfrac{x-6}{6} = -1$

16. $-3 = \dfrac{x-1}{5}$

17. $\dfrac{2-x}{3} = -4$

18. $-2 = \dfrac{1-x}{7}$

Example 4 Solve $28 = 9a + 5a$.

Solution

$$28 = 9a + 5a \quad \text{Combine } 9a \text{ and } 5a.$$
$$28 = 14a$$
$$\dfrac{28}{14} = \dfrac{14a}{14} \quad \text{Divide each side by 14.}$$
$$2 = a \quad \text{The solution set is } \{2\}.$$

Solve.

19. $4w - w = -12$

20. $20 = 2a + 3a$

21. $y - 4y = -18$

22. $5t + 3t = -16$

23. $-7v + 3v = -12$

24. $24 = -3n + 9n$

Example 5 Solve $3(y + 2) - 1 = 11$.

Solution

$$3(y + 2) - 1 = 11 \quad \left\{ \begin{array}{l} \text{Use the distributive property} \\ \text{to rewrite the left side.} \end{array} \right.$$
$$3y + 6 - 1 = 11$$
$$3y + 5 = 11$$
$$3y + 5 - 5 = 11 - 5 \quad \text{Subtract 5 from each side.}$$
$$3y = 6$$
$$\dfrac{3y}{3} = \dfrac{6}{3} \quad \text{Divide each side by 3.}$$
$$y = 2 \quad \text{The solution set is } \{2\}.$$

Solve.

25. $2(x - 1) = 16$

26. $3(y - 5) = 12$

27. $20 = 4(x + 3)$

28. $5(n + 2) - 3 = -18$

29. $6(x - 4) + 5 = 11$

30. $-3 = 7(h - 2) + 11$

Mixed Review Exercises

Solve.

1. $\dfrac{1}{4}x = -17$

2. $\dfrac{x}{6} = \dfrac{2}{3}$

3. $\dfrac{1}{4}x = 2\dfrac{1}{4}$

4. $-4 + x = -1$

5. $x + 7 = 16$

6. $30 = y + 12$

7. $-10 + x = -18$

8. $24 - x = 26$

9. $0.5x = -5$

10. $3.2 = n + 3$

11. $0 = 5x$

12. $14y = 280$

3–4 Using Equations to Solve Problems

Objective: To use the five-step plan to solve word problems.

Example 1   The sum of 25 and twice a number is 93. Find the number.

Solution

Steps 1, 2   Let n = the number. Then $2n$ = twice the number.

Step 3   The sum of 25 and twice a number is 93.
$$25 + 2n = 93$$

Step 4   Solve. $25 - 25 + 2n = 93 - 25$
$$2n = 68$$
$$n = 34$$

Step 5   *Check in the words of the problem:*   Is the sum of 25 and twice 34 equal to 93?
$$25 + 2(34) \stackrel{?}{=} 93$$
$$25 + 68 \stackrel{?}{=} 93$$
$$93 = 93 \ \checkmark \qquad \text{The number is 34.}$$

Solve each problem using the five-step plan to help you.

1. The sum of 17 and twice a number is 87. Find the number.

2. The sum of 8 and three times a number is 128. Find the number.

3. Seven more than twice a number is 175. Find the number.

4. Four less than half a number is 15. Find the number.

5. When one half of a number is decreased by 13, the result is 62. Find the number.

6. Six less than two thirds of a number is 18. Find the number.

Example 2   Find four consecutive even integers whose sum is 44.

Solution

Steps 1, 2   Let n = the first integer. Then $n + 2$ = the second integer, $n + 4$ = the third integer, and $n + 6$ = the fourth integer.

Step 3   The sum of the four consecutive even integers is 44.
$$n + (n + 2) + (n + 4) + (n + 6) = 44$$

Step 4   Solve. $4n + 12 = 44$ $\begin{cases} \text{If you're careful, you can subtract 12} \\ \text{from each side in your head.} \end{cases}$
$$4n = 32$$
$$n = 8 \quad \leftarrow \text{the first integer}$$
$$n + 2 = 10 \quad \leftarrow \text{the second integer}$$
$$n + 4 = 12 \quad \leftarrow \text{the third integer}$$
$$n + 6 = 14 \quad \leftarrow \text{the fourth integer}$$

Step 5   *Check:* $8 + 10 + 12 + 14 \stackrel{?}{=} 44$
$$44 = 44 \ \checkmark \qquad \text{The numbers are 8, 10, 12, and 14.}$$

3–4 Using Equations to Solve Problems (continued)

Solve each problem using the five-step plan to help you.

7. Find three consecutive integers whose sum is 138.

8. Find three consecutive odd integers whose sum is 87.

9. Find three consecutive even integers whose sum is 150.

10. Find four consecutive odd integers whose sum is 144.

11. Find five consecutive integers whose sum is 160.

12. Otto has $140. If he saves $2.50 per week, how long will it take him to have $200?

Example 3 The length of a rectangle is 9 cm more than the width. The perimeter is 78 cm. Find the length and the width.

Solution

Step 1 Draw a diagram to help you understand the problem.

Step 2 Let x = the width. Then $x + 9$ = the length.

Step 3
$$\text{perimeter} = 78$$
$$x + (x + 9) + x + (x + 9) = 78$$

Step 4 Solve.
$$4x + 18 = 78$$
$$4x = 60$$
$$x = 15 \quad \text{and} \quad x + 9 = 24$$

Step 5 *Check:* Is the sum of the lengths of the sides 78 cm?
$$15 + 24 + 15 + 24 \overset{?}{=} 78$$
$$78 = 78 \ \checkmark \qquad \text{The width is 15 cm. The length is 24 cm.}$$

Solve each problem using the five-step plan. Draw a diagram to help you.

3. The length of a rectangle is 11 cm more than the width. The perimeter is 90 cm. Find the length and width of the rectangle.

4. The width of a rectangle is 12 cm less than the length. The perimeter is 120 cm. Find the length and width of the rectangle.

5. The perimeter of a rectangle is 232 cm and the width is 56 cm. Find the length of the rectangle.

6. The perimeter of a rectangle is 340 cm and the length is 90 cm. Find the width of the rectangle.

Mixed Review Exercises

Solve.

1. $-3 + y = 2$

2. $x - 1.2 = 6$

3. $y + 6 = 15$

4. $\frac{2}{3}y = 6$

5. $-15 = \frac{c}{2}$

6. $-\frac{1}{5}x = 12$

7. $31 = y - 9$

8. $x - 15 = 16$

9. $0.25y = 8$

10. $3y + 2 = 17$

11. $2x - 3 = 15$

12. $3(a - 1) + 5 = 32$

3–5 Equations with the Variable on Both Sides

Objective: To solve equations with the variable on both sides.

Example 1　　Solve $5x = 2x + 15$.

Solution　　$5x - 2x = 2x + 15 - 2x$　　Subtract $2x$　　　　$Check:$　$5(5) \overset{?}{=} 2(5) + 15$
　　　　　　　　　　$3x = 15$　　　　　　　from each side.　　　　　　　　$25 \overset{?}{=} 10 + 15$
　　　　　　　　　　　$x = 5$　　　　　　　　　　　　　　　　　　　　　　　$25 = 25 \checkmark$

　　　　　　　The solution set is $\{5\}$.

Example 2　　Solve $4x = 30 - x$.

Solution　　　$4x + x = 30 - x + x$　　Add x to each side.
　　　　　　　　　　$5x = 30$
　　　　　　　　　　　$x = 6$　　　　　The solution set is $\{6\}$.

Solve.

1. $5n = 3n + 8$　　　　**2.** $7a = 2a + 30$　　　　**3.** $y = 20 - 3y$　　　　**4.** $3b = 80 - 5b$

5. $10n = 36 - 2n$　　　**6.** $2x = 20 - 8x$　　　　**7.** $21a = 56 + 7a$　　　**8.** $30 + 6x = 12x$

9. $-9a = -12a - 45$　**10.** $33c + 60 = 21c$　**11.** $72 - 4n = -22n$　**12.** $-11a = -12a -$

Example 3　　Solve $\frac{2}{5}x + 12 = x$.

Solution　　　$\frac{2}{5}x + 12 - \frac{2}{5}x = x - \frac{2}{5}x$　　Subtract $\frac{2}{5}x$ from each side.

　　　　　　　　　　　　　$12 = \frac{5}{5}x - \frac{2}{5}x$　　Rewrite $1x$ as $\frac{5}{5}x$.

　　　　　　　　　　　　　$12 = \frac{3}{5}x$

　　　　　　　　$\frac{5}{3} \cdot \frac{12}{1} = \frac{5}{3}\left(\frac{3}{5}x\right)$　　Multiply each side by $\frac{5}{3}$, the reciprocal of $\frac{3}{5}$.

　　　　　　　　　　　　　$20 = x$　　The solution set is $\{20\}$.

Example 4　　Solve $\frac{6 + x}{3} = x$.

Solution　　　$3\left(\frac{6 + x}{3}\right) = 3 \cdot x$　　Multiply each side by 3, the reciprocal of $\frac{1}{3}$.

　　　　　　　　　　$6 + x = 3x$
　　　　　　　$6 + x - x = 3x - x$　　Subtract x from each side.
　　　　　　　　　　　　$6 = 2x$
　　　　　　　　　　　　$3 = x$　　　The solution set is $\{3\}$.

Study Guide, ALGEBRA, Structure and Method, Book 1

–5 Equations with the Variable on Both Sides *(continued)*

olve.

3. $\frac{2}{3}x - 5 = x$ **14.** $\frac{3}{4}x - 8 = x$ **15.** $x = \frac{1}{2}x + 7$ **16.** $x = \frac{4}{5}x - 9$

7. $\frac{x-2}{3} = x$ **18.** $\frac{3+y}{4} = y$ **19.** $y = \frac{7-2y}{5}$ **20.** $x = \frac{9+x}{4}$

ocabulary

Empty set or null set The set with no members.

Identity An equation that is true for every value of the variable(s).

ymbol ϕ (empty set, or the null set)

AUTION   An equation may have no solution, or it may be satisfied by
every real number.

Example 5 Solve: **a.** $5(a - 2) - 3 = 3a + 7 + 2a$ **b.** $\frac{1}{3}(24x - 15) = 8x - 5$

Solution **a.** $5a - 10 - 3 = 5a + 7$ **b.** $8x - 5 = 8x - 5 \leftarrow$ **Identity**
$\qquad\qquad 5a - 13 = 5a + 7$ An identity is true for every
$\qquad\qquad\qquad -13 = 7 \leftarrow$ **False** value of the variable.

The equation has *no solution*. The solution set is {real numbers}.

---

olve each equation. If the equation is an identity or if it has no
lution, write *identity* or *no solution*.

. . $2(x - 3) = 5x$ **22.** $4(y - 5) = 9y$ **23.** $3n = 6(3 - n)$

. . $-3m = 5(2 - m)$ **25.** $2(a - 1) = 2a + 3$ **26.** $\frac{1}{4}(28x - 8) = 7x - 2$

. $\frac{1}{3}(3x - 3) + 2 = 2x$ **28.** $4(a - 1) - 5 = 3a + 7$ **29.** $3(5 + y) - y = 2y + 15$

. $4a + 7 + a = 3(a - 1)$ **31.** $\frac{3n - 15}{4} = 2n$ **32.** $\frac{2n - 9}{2} = n$

lixed Review Exercises

mplify.

. . $3 + \left(-\frac{1}{3}\right) + \left(-\frac{5}{3}\right)$ **2.** $-2\frac{3}{4} + 1\frac{1}{4}$ **3.** $-115 - (-10)$

. . $15x + (-3x) - 2$ **5.** $-4y + 5 + 18y + 23$ **6.** $6(-2)(-5)(-4)$

olve.

. $-2 - x = 5$ **8.** $4 + (1 + k) = 2$ **9.** $3x = -276$

. $\frac{1}{2}x = 3\frac{1}{2}$ **11.** $\frac{x}{6} = 7$ **12.** $-10\frac{2}{3} = -\frac{1}{3}x$

3–6 Problem Solving: Using Charts

Objective: To organize the facts of a problem in a chart.

Example 1   Organize the given information in a chart: In game 1 Jesse scored twice as many points as Ramon. In game 2 Jesse scored six fewer points than he did in game 1. In game 2 Ramon scored eight more points than he did in game 1.

Solution

	Game 1 points	Game 2 points
Jesse	$2n$	$2n - 6$
Ramon	n	$n + 8$

Example 2   Solve the problem using the two given facts:
Find the number of Calories in a banana and in a peach.
(1) A banana contains 65 Calories more than a peach.
(2) Ten peaches have 50 fewer Calories than 4 bananas.

Solution

Step 1   The problem asks for the number of Calories in a banana and in a peach.

Step 2   Let p = the number of Calories in a peach.
Then $p + 65$ = the number of Calories in a banana.

	Calories per fruit $\times$	Number of fruit $=$	Total Calories
Peach	p	10	$10p$
Banana	$p + 65$	4	$4(65 + p)$

Step 3   Calories in 10 peaches = Calories in 4 bananas $- 50$
$$10p = 4(p + 65) - 50$$

Step 4   Solve.
$$10p = 4p + 260 - 50$$
$$6p = 210$$
$$p = 35 \text{ and } p + 65 = 100$$

Step 5   *Check:* (1) 100 Calories is 65 more than 35 Calories. (2) Ten peaches have $10 \cdot 35$, or 350, Calories and four bananas have $4 \cdot 100$, or 400, Calories. $350 = 400 - 50$
There are 35 Calories in a peach and 100 Calories in a banana.

Solve each problem using the two given facts. If a chart is given, complete the chart to help you solve the problem.

1. Find the number of full 8 hour shifts that Cleo worked last month.
   (1) He worked three times as many 8 hour shifts as 6 hour shifts.
   (2) He worked a total of 180 hours.

	Hours per Shift $\times$	No. of Shifts $=$	Total hours worked
6 h shift	?	x	?
8 h shift	?	?	?

NAME _Taylor Kurtz_____   DATE _____

–6 Problem Solving: Using Charts (continued)

. Find the total weight of the boxes of cheddar cheese in a shipment of
3 lb boxes of cheddar cheese and 2 lb boxes of Swiss cheese. *330 lb*
(1) There were 20 fewer 2 lb boxes of Swiss cheese than 3 lb boxes of cheddar cheese.
(2) The total weight of the shipment was 510 lb.

	Weight per box ×	Number of boxes =	Total weight
Cheddar	? *3*	x	*330* ?
Swiss	? *2*	? *x − 20*	*180* ?

3x
2x − 40
3x + 2x − 40 = 510
+40   +40
5x = 550 ⟶ $\frac{5x}{5} = \frac{550}{5}$

. Find the number of 20-ride tickets sold. *20 - twenty ride tickets*
(1) Twenty times as many 8-ride tickets as 20-ride tickets were sold.
(2) The total number of tickets represented 3600 rides.

	Rides per ticket ×	Number of tickets sold =	Total rides
20-ride tickets	? *10*	n	*20n* ?
8-ride tickets	? *8*	*20n* ?	*160n* ?

$20n + 160n =$
$\frac{180n}{180} = \frac{3600}{180}$

Find the amount of time Maurice spent taking bowling lessons. *24 hours*
(1) He took three times as many 2 h bowling lessons as he did 1 h tennis lessons.
(2) He spent a total of 28 h taking bowling lessons and tennis lessons.

	Hours per lesson ×	Number of lessons =	Total time
Bowling	? *2*	*3x* ?	*6x* ?
Tennis	? *1*	*x* ?	*x* ?

$6x + x = 28$
$7x = 28$
$\frac{7x}{7} = \frac{28}{7}$
$x = 4$

Find the number of Calories in a grapefruit and an orange. *50 cal.   65 cal.*
(1) An orange has 15 more Calories than a grapefruit. *o = 15 + g*
(2) Twenty oranges and ten grapefruit have 1800 Calories. $20(15 + g) + 10g = 1800$

Find the number of Calories in a honeydew and in a cantaloupe.
(1) A honeydew has 20 more Calories than a cantaloupe. *h = 20 + c*
(2) Six honeydew and three cantaloupes have 750 Calories.

$300 + 30g = 1800$
$\frac{30g}{30} = \frac{1500}{30}$

$6(20 + c) + 3c = 750$
$120 + 9c = 750$
$-120 \quad -120$
$9c = 630$
$c = 70$

Mixed Review Exercises

Solve.

. $15x = 360$ *X = 24*

2. $6 = \frac{3}{5}x$ *X = 10*

3. $9z - 5z = 0$ *Z = 0*

. $165 = 3x$ *X = 55*

5. $6y + 5 = 35$ *y = 5*

6. $-10 + 3y = -28$ *y = 6*

. $4x - x = 21$ *X = 7*

8. $3(x + 2) = 4x$ *X = *

9. $6x - 7 = 2x + 41$ *X = 28*

. $21 - x = 1 - 6x$ *y = -4*

11. $-x = 3x - 52$ *X = 13*

12. $5(y + 1) + 3 = 3y - 20$ *y = 6*

Solve.
. $15x = 360$ *X = 24*
2. $6 = \frac{3}{5}x$ *X = 10*
3. $9z - 5z = 0$ *Z = 0*
. $165 = 3x$ *X = 55*
5. $6y + 5 = 35$ *y = 5*
6. $-10 + 3y = -28$ *y = 6*
. $4x - x = 21$ *X = 7*
8. $3(x + 2) = 4x$
9. $6x - 7 = 2x + 41$ *X = 28*
. $21 - x = 1 - 6x$ *y = -4*
11. $-x = 3x - 52$ *X = 13*
12. $5(y + 1) + 3 = 3y - 20$ *y = 6*

3–7 Cost, Income, and Value Problems

Objective: To solve problems involving cost, income, and value.

Formulas

Cost = number of items × price per item

Income = hours worked × wage per hour

Total value = number of items × value per item

Example 1   Tickets for a concert cost $8 for adults and $4 for students. A total of 920 tickets worth $5760 were sold. How many adult tickets were sold?

Solution

Step 1   The problem asks for the number of adult tickets sold.

Step 2   Let x = the number of adult tickets sold.
Then $920 - x$ = the number of student tickets sold.
Make a chart.

	Number	× Price per ticket =	Cost
Adult	x	8	$8x$
Student	$920 - x$	4	$4(920 - x)$

Step 3   The only fact not recorded in the chart is that the total cost of the tickets was $5760. Write an equation using this fact.

Adult ticket cost + Student ticket cost = 5760
$$8x + 4(920 - x) = 5760$$

Step 4
$$8x + 4(920 - x) = 5760$$
$$8x + 3680 - 4x = 5760$$
$$4x + 3680 = 5760$$
$$4x = 2080$$
$$x = 520 \longleftarrow \text{adult tickets}$$
$$920 - x = 400 \longleftarrow \text{student tickets}$$

Step 5   *Check:*   520 adult tickets at $8 each cost $4160.
400 student tickets at $4 each cost $1600.

The total number of tickets is 520 + 400, or 920. √
The total cost of the tickets is $4160 + $1600, or $5760. √

520 adult tickets were sold.

Solve. Complete the chart first.

1. Forty students bought caps at the baseball game. Plain caps cost $4 each and deluxe ones cost $6 each. If the total bill was $236, how many students bought the deluxe cap?

	Number ×	Price =	Cost
Deluxe	d	6	$6d$
Plain	40 - ?d	4	160 - 4

(236)

3-7 Cost, Income, and Value Problems (continued)

Solve. Complete the chart first.

2. Adult tickets for the game cost $6 each and student tickets cost $3 each. A total of 1040 tickets worth $5400 were sold. How many student tickets were sold?

	Number	×	Price	=	Cost
Adult	?		?		?
Student	s		?		?

3. A collection of 60 dimes and nickels is worth $4.80. How many dimes are there? (*Hint:* In your equation, use 480¢, instead of $4.80.)

	Number	×	Value of coin	=	Total value
Dimes	d		?		?
Nickels	?		?		?

4. A collection of 54 dimes and nickels is worth $3.80. How many nickels are there? (*Hint:* In your equation, use 380¢ instead of $3.80.)

	Number	×	Value of coin	=	Total value
Dimes	?		?		?
Nickels	n		?		?

5. Henry paid $.80 for each bag of peanuts. He sold all but 20 of them for $1.50 and made a profit of $54. How many bags did he buy? (*Hint:* Profit = selling price − buying price.)

	Number	×	Price (¢)	=	Cost (¢)
Bought	b		?		?
Sold	?		?		?

6. Paula paid $4 for each stadium cushion. She sold all but 12 of them for $8 each and made a profit of $400. How many cushions did she buy? (*Hint:* Profit = selling price − buying price.)

	Number	×	Price (¢)	=	Cost (¢)
Bought	b		?		?
Sold	?		?		?

Solve. Make and complete a chart first.

7. I have three times as many dimes as quarters. If the coins are worth $6.60, how many quarters are there?

8. I have 12 more nickels than quarters. If the coins are worth $5.40, how many nickels are there?

Mixed Review Exercises

Simplify.

1. $\dfrac{30 \div 5 + 2}{13 - 5}$

2. $24 \div \dfrac{1}{4}$

3. $\dfrac{1}{4}(28y - 12) + 6$

4. $(-5)(4)(-2)$

5. $3(2x + 5) + 4(-x)$

6. $6(x - y) + 5(2y + x)$

Evaluate if $a = 2$, $b = 3$, and $c = 8$.

7. $\dfrac{3a + b}{c - 5}$

8. $\dfrac{bc}{2a}$

9. $2(c - a) - b \div 3$

3–8 Proof in Algebra

Objective: To prove statements in algebra.

Vocabulary

Theorem A statement that is shown to be true using a logically developed argument.

Proof Logical reasoning that uses given facts, definitions, properties, and other already proved theorems to show that a theorem is true. You may refer to the Chapter Summary, on page 88 of your textbook, for listings of properties and theorems that you can use as reasons in your proofs.

Example 1 *Prove:* If $a + b = 0$, then $b = -a$.

Proof

	Statements	Reasons
1.	$a + b = 0$	1. Given
2.	$-a + (a + b) = -a + 0$	2. Addition property of equality
3.	$(-a + a) + b = -a + 0$	3. Associative property of addition
4.	$0 + b = -a + 0$	4. Property of opposites
5.	$b = -a$	5. Identity property of addition

Example 2 *Prove:* $a \cdot 0 = 0$

Proof

	Statements	Reasons
1.	$0 = 0 + 0$	1. Identity property of addition
2.	$a \cdot 0 = a(0 + 0)$	2. Multiplication property of equality
3.	$a \cdot 0 = a \cdot 0 + a \cdot 0$	3. Distributive property
4.	$a \cdot 0 = a \cdot 0 + 0$	4. Identity property of addition
5.	$a \cdot 0 + a \cdot 0 = a \cdot 0 + 0$	5. Substitution principle
6.	$a \cdot 0 = 0$	6. Subtraction property of equality

Write the missing reasons. Assume that each variable represents any real number.

1. *Prove:* For all real numbers a and b, $-b + (a + b) = a$.

Proof:

	Statements	Reasons
1.	$-b + (a + b) = -b + (b + a)$	1. ___?___
2.	$= (-b + b) + a$	2. ___?___
3.	$= 0 + a$	3. ___?___
4.	$= a$	4. ___?___

–8 Proof in Algebra *(continued)*

Write the missing reasons. Assume that each variable represents any real number.

2. *Prove:* For all real numbers a and b, $-a(b + c) = -ab - ac$.

Proof:

Statements	Reasons
1. $-a(b + c) = (-a)b + (-a)c$	1. ?
2. $\quad\quad = -ab + (-ac)$	2. ?
3. $\quad\quad = -ab - ac$	3. ?

3. *Prove:* For all real numbers a and b, $b \neq 0$, $(ab) \div b = a$.

Proof:

Statements	Reasons
1. $(ab) \div b = (ab) \cdot \dfrac{1}{b}$	1. ?
2. $\quad\quad = a\left(b \cdot \dfrac{1}{b}\right)$	2. ?
3. $\quad\quad = a \cdot 1$	3. ?
4. $\quad\quad = a$	4. ?

4. *Prove:* For all real numbers a and b, $b \neq 0$, $\dfrac{a + b}{b} = 1 + \dfrac{a}{b}$.

Proof:

Statements	Reasons
1. $\dfrac{a + b}{b} = (a + b) \cdot \dfrac{1}{b}$	1. ?
2. $\quad\quad = (a) \cdot \dfrac{1}{b} + (b) \cdot \dfrac{1}{b}$	2. ?
3. $\quad\quad = (a) \cdot \dfrac{1}{b} + 1$	3. ?
4. $\quad\quad = \dfrac{a}{b} + 1$	4. ?
5. $\quad\quad = 1 + \dfrac{a}{b}$	5. ?

Mixed Review Exercises

Simplify.

1. $8(10 - 3) \div 4 + 2$ **2.** $-3(-12 + 4)$ **3.** $-6x - x + 9x$

4. $-\dfrac{1}{4}(8 + 4a)$ **5.** $\dfrac{1}{3}(3b + 9c)$ **6.** $12 \div \dfrac{1}{3}$

Evaluate if $a = 4$, $b = 3$, $c = 2$, and $x = 12$.

7. $a(x - b)$ **8.** $2|b - x|$ **9.** $x - |c - a|$

10. $\dfrac{x + 2a}{|1 - b|c}$ **11.** $\dfrac{4a + x + 7}{b + c}$ **12.** $\dfrac{1}{3}(x - b) + a$

4 Polynomials

4–1 Exponents

Objective: To write and simplify expressions involving exponents.

Vocabulary

Power of a number A product of equal factors. For example, $2 \times 2 \times 2$, or 2^3, is the third power of 2.

Base of a power The number that is used as a factor. For example, 2 is the base in 2^3.

Exponent In a power, the number that indicates how many times the base is used as a factor. For example, 3 is the exponent in 2^3.

Exponential form The expression 2^3 is the exponential form of $2 \cdot 2 \cdot 2$.

CAUTION 1 Be careful when an expression contains both parentheses and exponents.
$(3y)^2$ means $(3y)(3y)$.   2 is the exponent of the base $3y$.
$3y^2$ means $3 \cdot y \cdot y$.   2 is the exponent of the base y.

CAUTION 2 Follow the correct order when simplifying expressions.
1. First simplify expressions within grouping symbols.
2. Then simplify powers.
3. Then simplify products and quotients in order from left to right.
4. Then simplify sums and differences in order from left to right.

Example 1	Write each expression in exponential form.

Example 1   Write each expression in exponential form.
a. $5 \cdot 5 \cdot 5$ b. $a \cdot a \cdot a \cdot a \cdot a$ c. $-2 \cdot x \cdot y \cdot 5 \cdot x \cdot x$

Solution   a. 5^3 b. a^5 c. $-10x^3y$

Write each expression in exponential form.

1. $x \cdot x \cdot x \cdot x$
2. $m \cdot m$
3. $3 \cdot t \cdot t \cdot t \cdot t$
4. $c \cdot e \cdot 2 \cdot c \cdot e$
5. $-4 \cdot z \cdot z \cdot z$
6. $y \cdot y \cdot (-2)$
7. $-2 \cdot x \cdot x \cdot 3 \cdot x$
8. $5 \cdot n \cdot (-3)$
9. $a \cdot a \cdot a \cdot b \cdot b$
10. $c \cdot c \cdot d \cdot d \cdot d \cdot c$
11. $m \cdot 6 \cdot n \cdot m$
12. $u \cdot v \cdot u \cdot v \cdot 8$

Example 2   Evaluate x^3 if $x = -2$.

Solution $x^3 = (-2)^3$ Replace x with -2.
$= (-2)(-2)(-2)$ Simplify.
$= -8$

---

13. Evaluate x^4 if $x = -2$.
14. Evaluate x^3 if $x = -3$.
15. Evaluate y^5 if $y = -1$.
16. Evaluate x^2 if $x = -5$.

4–1 Exponents (continued)

Example 3   Find the area of the rectangle.

4x
x

Solution   Area = length × width
$= 4x \cdot x$
$= 4x^2$

Find the area of each rectangle.

17.
5y
3

18.
3x
2x

19.
8z
3z

Example 4   Simplify: **a.** -2^4 **b.** $(-2)^4$ **c.** $(1 + 2)^3$ **d.** $1 + 2^3$

Solution   **a.** $-2^4 = -(2 \cdot 2 \cdot 2 \cdot 2) = -16$ **b.** $(-2)^4 = (-2)(-2)(-2)(-2) = 16$
c. $(1 + 2)^3 = 3^3 = 27$ **d.** $1 + 2^3 = 1 + 2 \cdot 2 \cdot 2 = 1 + 8 = 9$

Simplify.

20. a. 3^4
  b. 4^3

21. a. -3^2
  b. $(-3)^2$

22. a. -2^3
  b. $(-2)^3$

23. a. $2 \cdot 4^2$
  b. $(2 \cdot 4)^2$

24. a. $5 - 2^2$
  b. $(5 - 2)^2$

25. a. $2 - 5^2$
  b. $(2 - 5)^2$

26. a. $5 - 3^2$
  b. $(5 - 3)^2$

27. a. $2 \cdot 5 - 3^2$
  b. $(2 \cdot 5 - 3)^2$

Example 5   Evaluate $(2a - b)^2$ if $a = 2$ and $b = -3$.

Solution $(2a - b)^2 = [2 \cdot 2 - (-3)]^2$ Replace a with 2 and b with -3.
$= [4 + 3]^2$ Simplify the expression within the brackets.
$= 7^2$
$= 49$

Evaluate each expression if $x = 2$ and $y = -1$.

28. a. $2x + y^2$
  b. $(2x + y)^2$

29. a. $2 + xy^2$
  b. $(2 + xy)^2$

30. a. $2x + y^3$
  b. $(2x + y)^3$

31. a. $(x + 2y)^3$
  b. $x^3 + 2y^3$

32. a. $2x - y^2$
  b. $(2x - y)^2$

33. a. $2 - xy^2$
  b. $(2 - xy)^2$

34. a. $2x - y^3$
  b. $(2x - y)^3$

35. a. $(x - 2y)^3$
  b. $x^3 - 2y^3$

Mixed Review Exercises

Solve.

1. $-6x = 42$

2. $2(n - 3) = 24$

3. $24 = -8x$

4. $-n + 6 = 8$

5. $x - 5 = -3$

6. $-y + 10 = 6$

7. $-\frac{1}{2}(x + 3) = 4$

8. $\frac{1}{3}x = 10$

NAME _____ DATE _____

4–2 Adding and Subtracting Polynomials

Objective: To add and subtract polynomials.

Vocabulary

Monomial An expression that is either a numeral, a variable, or the product of a numeral and one or more variables. For example: 13, m, $8c$, $2xy$, $5p^2q$.

Coefficient In the monomial $9x^2y^3$, 9 is the coefficient, or numerical coefficient.

Similar, or like, terms Two monomials that are exactly alike or the same except for their numerical coefficient. For example, $-3xy$ and $7xy$ are similar.

Constant monomial or constant A numeral such as 7.

Polynomial A sum of monomials. For example, $x^2 + 3x + y^2 + 2$.

Binomial A polynomial of only two terms. For example, $2x + 5$.

Trinomial A polynomial of only three terms. For example, $a^2 + 2ab + b^2$.

Simplified form, or simplest form, of a polynomial A polynomial which has no two of its terms similar.

CAUTION When a monomial does not have a written numerical coefficient, remember that its coefficient is 1. For example, $x^6y^2 = 1x^6y^2$.

Example 1 Simplify $-5x^3 + 2x^2 + x^2 + 7x^3 - 4$.

Solution
$$-5x^3 + 2x^2 + x^2 + 7x^3 - 4 = (-5x^3 + 7x^3) + (2x^2 + x^2) - 4$$
$$= (-5 + 7)x^3 + (2 + 1)x^2 - 4$$
$$= 2x^3 + 3x^2 - 4$$

Simplify.

1. $2x - y + 3x - 2y$

2. $7m - 5n - 2m + n$

3. $4x^2 - 3x - 2x^2 + 7x - 2$

4. $n^2 - 3n - 5n^2 + 7n + 6n^2$

5. $a^2 + 2ab - 5ab + 4a^2$

6. $x^2y - y^3 - 8x^2y + 5y^3$

7. $a^2b - 2ab^2 + 5a^3 - 3a^2b$

8. $-6xy^2 + 5x^2y - x^3 + xy^2 + 3x^3 - 2x^2y$

Example 2 Add $2x^2 + 5xy - 6y^2$ and $8x^2 + 6xy + y^2$.

Solution 1 First group similar terms and then combine them.
$$(2x^2 + 5xy - 6y^2) + (8x^2 + 6xy + y^2) = (2x^2 + 8x^2) + (5xy + 6xy) + (-6y^2 + y^2)$$
$$= 10x^2 + 11xy - 5y^2$$

Solution 2
$$\begin{array}{l} 2x^2 + 5xy - 6y^2 \\ 8x^2 + 6xy + y^2 \\ \hline 10x^2 + 11xy - 5y^2 \end{array}$$

You can also align similar terms vertically and add.

Study Guide, ALGEBRA, Structure and Method, Book 1
Copyright © by Houghton Mifflin Company. All rights reserved.

–2 Adding and Subtracting Polynomials (continued)

Vocabulary

Degree of a variable in a monomial The number of times that the variable occurs as a factor in a monomial. For example, in $7x^3y$, the degree of x is 3, and the degree of y is 1.

Degree of a monomial The sum of the degrees of its variables. For example, the degree of $8x^2y^3$ is 5. The degree of any nonzero constant monomial, such as 10, is 0.

Degree of a polynomial The greatest of the degrees of its terms after it has been simplified. For example, the degree of $-5x^3 + 2x^2 + x^2 + 5x^3 - 4$ is 2, since it can be simplified to $3x^2 - 4$.

Add.

9. $\begin{array}{l} 3a - 1 \\ \underline{4a + 3} \end{array}$ 10. $\begin{array}{l} 4n + 2 \\ \underline{-2n - 5} \end{array}$ 11. $\begin{array}{l} 2x - 3y \\ \underline{-2x + 6y} \end{array}$ 12. $\begin{array}{l} 5n - 2p \\ \underline{-3n + 5p} \end{array}$

13. $\begin{array}{l} 4x - 5y + 3 \\ \underline{-2x + 7y + 7} \end{array}$ 14. $\begin{array}{l} 2a - 3b - 6 \\ \underline{3a - \ b + 8} \end{array}$ 15. $\begin{array}{l} 6x^2 - 3x + 2 \\ \underline{2x^2 + \ x - 5} \end{array}$ 16. $\begin{array}{l} 5 - 2n - 6n^2 \\ \underline{-3 + \ n - 2n^2} \end{array}$

17. $\begin{array}{l} 4c^2 - 3cd - 5d^2 \\ \underline{-c^2 + 6cd - 2d^2} \end{array}$ 18. $\begin{array}{l} 6a^2 - 2ab \\ \underline{-2a^2 + 5ab - b^2} \end{array}$ 19. $\begin{array}{l} 3x - 2y - 5z + 1 \\ 2x + \ y - 3z \\ \underline{3y + \ z + 3} \end{array}$ 20. $\begin{array}{l} 6a - 2b \quad\ \ + 4 \\ 3a \qquad\quad - 5c - 1 \\ \underline{-a - \ b + 6c + 5} \end{array}$

Example 3   Subtract $-x^2 + 5xy + 6y^2 - 3$ from $3x^2 - 6xy - 2y^2 - 5$.

Solution 1   Add the opposite of $(-x^2 + 5xy + 6y^2 - 3)$ to $3x^2 - 6xy - 2y^2 - 5$.
$(3x^2 - \ 6xy - 2y^2 - 5) - (-x^2 + 5xy + 6y^2 - 3) =$
$(3x^2 - \ 6xy - 2y^2 - 5) + (\ x^2 - 5xy - 6y^2 + 3) = 4x^2 - 11xy - 8y^2 - 2$

Solution 2   You can also align similar terms vertically.

$\begin{array}{l} 3x^2 - 6xy - 2y^2 - 5 \\ \underline{-(-x^2 + 5xy + 6y^2 - 3)} \end{array}$ $\xrightarrow[\text{and add.}]{\begin{array}{c}\text{Change to the}\\ \text{opposite signs}\end{array}}$ $\begin{array}{l} 3x^2 - \ 6xy - 2y^2 - 5 \\ \underline{\ x^2 - \ 5xy - 6y^2 + 3} \\ 4x^2 - 11xy - 8y^2 - 2 \end{array}$

21–30. In Exercises 9–18, subtract the lower polynomial from the upper one.

Mixed Review Exercises

Simplify.

1. -2^3 2. $(-3)^2$ 3. $2^2 + 3^2$ 4. $(2 + 3)^2$

Solve.

5. $3(y + 2) - 2 = 2(4 - y)$ 6. $10 = 2(n + 3)$ 7. $4(x - 10) = 13 - 3(2x + 1)$

8. $-\dfrac{2}{5}(n + 3) = 10$ 9. $c - 2 = |1 - 8|$ 10. $\dfrac{3}{4}(2y - 6) = y - 7$

4–3 *Multiplying Monomials*

Objective: To multiply monomials.

Rule of Exponents for Products of Powers	Example
For all positive integers m and n: $$a^m \cdot a^n = a^{m+n}.$$ This means that when you multiply two powers having the same base, you add the exponents.	$$x^3 \cdot x^5 = x^{3+5} = x^8$$

CAUTION   Use the rule of exponents for products of powers only when the two powers to be multiplied have the *same base*. For example,

$$m^2 \cdot n^3 = m^2 n^3, \text{ not } mn^5$$

Example 1   Simplify: **a.** $x^2 \cdot x^5$ **b.** $c^6 \cdot c^3$ **c.** $a \cdot a^5$

Solution   **a.** $x^2 \cdot x^5 = x^{2+5} = x^7$

b. $c^6 \cdot c^3 = c^{6+3} = c^9$

c. $a \cdot a^5 = a^1 \cdot a^5 = a^{1+5} = a^6$

Simplify.

1. $x^3 \cdot x^6$

2. $c^7 \cdot c^8$

3. $a^9 \cdot a^{10}$

4. $x^2 \cdot x^3 \cdot x$

5. $n^2 \cdot n^2 \cdot n^3$

6. $c \cdot c^5 \cdot c^2$

7. $a^2 \cdot a^3 \cdot a^5$

8. $x^5 \cdot x^6 \cdot x^7$

9. $c^3 \cdot c^6 \cdot c^7$

10. $m^2 \cdot m^6 \cdot m^8$

11. $n^{10} \cdot n \cdot n^3$

12. $x \cdot x^9 \cdot x^{10}$

Example 2   Simplify $(2x^3)(4x^4)$.

Solution $(2x^3)(4x^4) = (2 \cdot 4)(x^3 \cdot x^4)$ {Use the commutative and associative

$= 8(x^{3+4})$ properties of multiplication.

$= 8x^7$ Use the rule of exponents for products of powers.

Simplify.

13. $(2a^4)(5a^3)$

14. $(4x^3)(3x^4)$

15. $(7m^5)(2m^6)$

16. $(5x^4)(3x^2)$

17. $(-2xy^2)(-3x^2y)$

18. $(4a^2b)(-3ab^3)$

19. $(3ab)(a^2b)(5b^2)$

20. $(6x^2y)(2xy^2)(3x)$

21. $(3cd^4)(-2c^2)(4cd^2)$

22. $(5a^3b^2)(-4a^2b^2)(-2ab^3)$

23. $(-x^2y^2)(3x^2y)(-4xy^3)$

24. $(-a^2b)(-5ab^3)(-b^2)$

–3 Multiplying Monomials (continued)

Example 3   Simplify $\left(\frac{10x^2y}{3}\right)\left(\frac{6x^3y^2}{5}\right)$.

Solution $\left(\frac{10x^2y}{3}\right)\left(\frac{6x^3y^2}{5}\right) = \left(\frac{\overset{2}{\cancel{10}}}{\underset{1}{\cancel{3}}} \cdot \frac{\overset{2}{\cancel{6}}}{\underset{1}{\cancel{5}}}\right)(x^2 \cdot x^3)(y \cdot y^2) = 4x^5y^3$

Simplify.

25. $\left(\frac{3}{4}r^2\right)\left(\frac{4}{3}t^2\right)$

26. $\left(\frac{6h^2k^3}{5}\right)\left(\frac{20hk^2}{3}\right)$

27. $(8a)\left(\frac{3}{4}a^2\right)$

28. $(12c^2)\left(-\frac{5}{6}cd^2\right)$

29. $\left(\frac{3a^2}{7}\right)(35a^5)$

30. $\left(\frac{8x^2y}{5}\right)\left(\frac{15xy^2}{16}\right)$

31. $\left(-\frac{5}{6}x^3\right)(3xy^2)(-y^2)$

32. $(3y^2)\left(\frac{2}{3}y^2\right)\left(\frac{1}{2}y\right)$

Example 4   Simplify $(2x^3)(-4x^2) + (5x^2)(3x^3)$.

Solution $(2x^3)(-4x^2) + (5x^2)(3x^3) = \underbrace{(2)(-4)(x^3 \cdot x^2)}_{} + \underbrace{(5 \cdot 3)(x^2 \cdot x^3)}_{}$

$= \underbrace{-8x^5 \quad + \quad 15x^5}_{}$

$= \quad 7x^5$

Simplify.

33. $(2x)(3x^3) + (5x^2)(4x^2)$

34. $(3x^5)(4x) - (5x^3)(2x^3)$

35. $(6x^2)(2x^3) + (3x)(5x^4)$

36. $(6x^5)(4x^2) - (2x^3)(12x^4)$

37. $(3a^4)(-2a^3) + (2a^2)(a^5)$

38. $(4y^2)(4y) - (5y^2)(3y)$

Mixed Review Exercises

Simplify.

1. $3 + 4^2$

2. $(3 + 4)^2$

3. $3a^2 + 5b^2 - a^2 - 2b^2$

4. $2 \cdot 5^2$

5. $(2 \cdot 5)^2$

6. $2x^2 - 3x + 4 + 5x + 3x^2$

Solve.

7. $3(y + 2) = 24$

8. $10z = 20 + 5z$

9. $6n - 12 = 2n$

10. $\frac{n}{4} + 2 = 5$

11. $3(x - 2) = 9$

12. $\frac{x}{3} - 1 = 2$

4–4 Powers of Monomials

Objective: To find powers of monomials.

Rules of Exponents	Examples
Rule of Exponents for a Power of a Power For all positive integers m and n: $\qquad (a^m)^n = a^{mn}.$ To find a power of a power, you multiply the exponents.	$(2^3)^4 = 2^{3 \cdot 4}$ $\qquad\quad = 2^{12}$
Rule of Exponents for a Power of a Product For every positive integer m: $\qquad (ab)^m = a^m b^m.$ To find a power of a product, you find the power of each factor and then multiply.	$(-2x)^5 = (-2)^5(x)^5$ $\qquad\qquad = -32x^5$

CAUTION $(x^7)^6 = x^{7 \cdot 6} = x^{42}$ but $x^7 \cdot x^6 = x^{7 + 6} = x^{13}$

Example 1   Simplify: **a.** $(x^2)^4$ **b.** $(u^3)^5$

Solution   Use the rule for a power of a power.

$\qquad$ **a.** $(x^2)^4 = x^{2 \cdot 4}$ $\qquad$ **b.** $(u^3)^5 = u^{3 \cdot 5}$
$\qquad\qquad\quad = x^8$ $\qquad\qquad\qquad\qquad = u^{15}$

Simplify.

1. $(a^2)^3$ $\qquad\qquad$ **2.** $(x^4)^3$ $\qquad\qquad$ **3.** $(t^5)^3$ $\qquad\qquad$ **4.** $(c^3)^3$

5. $(t^2)^3$ $\qquad\qquad$ **6.** $(x^5)^2$ $\qquad\qquad$ **7.** $(y^{10})^3$ $\qquad\qquad$ **8.** $(a^7)^8$

Example 2   Simplify: **a.** $(2x)^4$ **b.** $(-6k)^3$

Solution   Use the rule for a power of a product.

$\qquad$ **a.** $(2x)^4 = 2^4 \cdot x^4$ $\qquad$ **b.** $(-6k)^3 = (-6)^3 \cdot k^3$
$\qquad\qquad\quad = 16x^4$ $\qquad\qquad\qquad\qquad = -216k^3$

Simplify.

9. $(5a)^2$ $\qquad\qquad$ **10.** $(-6x)^2$ $\qquad\qquad$ **11.** $(-3t)^3$ $\qquad\qquad$ **12.** $(-4c)^2$

13. $(-5x)^3$ $\qquad\qquad$ **14.** $(-4t)^3$ $\qquad\qquad$ **15.** $(-2t)^4$ $\qquad\qquad$ **16.** $(6x)^3$

17. $(5x)^4$ $\qquad\qquad$ **18.** $(7n)^2$ $\qquad\qquad$ **19.** $\left(\frac{1}{2}a\right)^2$ $\qquad\qquad$ **20.** $\left(-\frac{1}{3}a\right)^3$

-4 Powers of Monomials (continued)

Example 3   Evaluate if $x = 3$:   **a.** $2x^3$ **b.** $(2x)^3$ **c.** $2^3 \cdot x^3$

Solution   **a.** $2x^3 = 2(3)^3$ **b.** $(2x)^3 = (2 \cdot 3)^3$ **c.** $2^3 \cdot x^3 = 2^3 \cdot 3^3$
$\qquad\qquad\quad = 2(27) \qquad\qquad\qquad\quad = 6^3 \qquad\qquad\qquad\qquad\quad = 8 \cdot 27$
$\qquad\qquad\quad = 54 \qquad\qquad\qquad\qquad\quad = 216 \qquad\qquad\qquad\qquad\quad = 216$

Evaluate if $x = 2$ and $y = 4$.

21. a. $2x^3$ **22. a.** $4y^2$ **23. a.** x^2y^3
  b. $(2x)^3$ **b.** $(4y)^2$ **b.** x^2y^2
  c. $2^3 \cdot x^3$ **c.** $4^2 \cdot y^2$ **c.** $(xy)^2$

24. a. xy^3 **25. a.** $3x^2$ **26. a.** $5x^2$
  b. $(xy)^3$ **b.** $(3x)^2$ **b.** $(5x)^2$
  c. $x^3 \cdot y^3$ **c.** $3^2 \cdot x^2$ **c.** $5^2 \cdot x^2$

27. a. xy^2 **28. a.** $2xy$ **29. a.** $6x^2 \div x$
  b. $(x^2y)^2$ **b.** $2x^2y$ **b.** $(6x)^2 \div x$
  c. x^3y **c.** $2xy^2$ **c.** $6(x^2 \div x)$

Example 4   Simplify $(-2x^2y^3)^4$.

Solution $(-2x^2y^3)^4 = (-2)^4(x^2)^4(y^3)^4$ $\left[\begin{array}{l}\text{First use the rule for a power of a product}\\ \text{and then use the rule for a power of a power.}\end{array}\right.$
$\qquad\qquad\qquad = 16x^8y^{12}$

Simplify.

30. $(3n^2)^3$ **31.** $(6b^4)^2$ **32.** $\left(\dfrac{1}{3}x^{10}\right)^3$

33. $\left(\dfrac{1}{2}x^2\right)^4$ **34.** $(2ab^2)^3$ **35.** $(-3x^2y^3)^3$

36. $(4x^3y^2)^3$ **37.** $(-2xy^2)^4$ **38.** $(5m^2n^4)^2$

Mixed Review Exercises

Simplify.

1. $(2a^2b)(3ab)(5ab^2)$ **2.** $(-xy^2)(2xy)(-3y)$

3. $(3x^2y^3)^4$ **4.** $\left(\dfrac{1}{3}t^2\right)\left(\dfrac{3}{4}t^3\right)$

5. $5c - 2a - 3c + a$ **6.** $(2x + 3y + 1) + (3x + 2y)$

7. $3 \cdot 5^2 + 3 \cdot 5$ **8.** $-3^2 \cdot 4$

9. $(3^3 + 5^2) \div 2^2$ **10.** $7 \cdot 3^2 + 6 \cdot 3 + 2$

11. $\left(\dfrac{5}{2}t^2\right)\left(\dfrac{1}{5}t^3\right)$ **12.** $(15mn^2)\left(\dfrac{1}{3}m^2\right)(4n)$

4–5 Multiplying Polynomials by Monomials

Objective: To multiply a polynomial by a monomial.

Example 1   Multiply: $x(x + 4)$

Solution 1 $x(x + 4) = x(x) + x(4)$
$= x^2 + 4x$

Solution 2 $\begin{array}{r} x + 4 \\ \underline{x} \\ x^2 + 4x \end{array}$ Multiply each term by x.

Multiply.

1. $3(x - 2)$ 2. $-2(x + 3)$ $c(c - 4)$ 4. $a(3 - 2a)$

5. $-2b(3 - 4b)$ 6. $-3c(4c + 1)$ 7. $5y(y + 6)$ 8. $-z(4 - 5z)$

Example 2   Multiply: $-2x(3x^2 - 2x + 1)$

Solution 1   Multiply each term of the polynomial $3x^2 - 2x + 1$ by the monomial $-2x$.

$-2x(3x^2 - 2x + 1) = -2x(3x^2) - 2x(-2x) - 2x(1)$
$= -6x^3 + 4x^2 - 2x$

Solution 2 $\begin{array}{r} 3x^2 - 2x + 1 \\ \underline{-2x} \\ -6x^3 + 4x^2 - 2x \end{array}$

Multiply.

9. $3x(x^2 - x - 2)$ 10. $-2x(x^2 - 4x + 5)$ 11. $-4x(2x^2 - 3x - 7)$

12. $5x^2(x^2 + x - 3)$ 13. $-6x^2(x^2 - x - 12)$ 14. $4x^3(x^2 - 3x - 6)$

15. $\begin{array}{r} 3a^2 - 4a - 6 \\ \underline{2a} \end{array}$ 16. $\begin{array}{r} 4a^2 - 5a - 7 \\ \underline{5a} \end{array}$ 17. $\begin{array}{r} 5x^2 - x - 3 \\ \underline{2x^2} \end{array}$ 18. $\begin{array}{r} 2k^2 - 3k - 5 \\ \underline{-4k^3} \end{array}$

Example 3   Multiply: $4x^2y(5x^2 - 3xy + 2y^2)$

Solution   Multiply each term of the polynomial by $4x^2y$.

$4x^2y(5x^2 - 3xy + 2y^2) = 4x^2y(5x^2) + 4x^2y(-3xy) + 4x^2y(2y^2)$
$= 20x^4y - 12x^3y^2 + 8x^2y^3$

Multiply.

19. $3x^2y(4x^2 - 5xy - 2y^2)$ 20. $xy^2(x^2 - 4xy - 5y^2)$

21. $-2xy(4x^2 - 3xy + y^2)$ 22. $\frac{1}{3}x^2y(6x^2 - 12xy + 9y^2)$

---

ultiply.

. $2xy^2(3x^2 - 7xy - 2y^2)$

24. $-4x^3y(x^2 - 3xy - 6y^2)$

. $5xy(2x^2 - 4xy + y^2)$

26. $\frac{1}{2}x^2y^2(6x^2 - 4xy - 8y^2)$

Example 4   Simplify $3n(n + 2) + n(5 - n)$.

Solution $3n(n + 2) + n(5 - n) = 3n(n) + 3n(2) + n(5) - n(n)$ $\left\{\begin{array}{l}\text{Use the distributive}\\ \text{property.}\end{array}\right.$
$ = 3n^2 + 6n + 5n - n^2$
$ = 2n^2 + 11n$ Combine similar terms.

---

mplify.

. $2x(x - 3) + 3x(x + 2)$

28. $4x(3 - 2x) + 5x(x - 1)$

. $5x^2(2x - 1) - 2x(3x^2 - 4x)$

30. $3y(4y^2 - 3y) - 2y^2(y + 1)$

. $2n^2(4n - 5) - 3n(2n^2 - 7n)$

32. $2x(5x^2 - 3x) - x^2(x + 6)$

Example 5   Solve $n(2 - 3n) + 3(n^2 - 4) = 0$.

Solution
$n(2 - 3n) + 3(n^2 - 4) = 0$ Use the distributive property.
$2n - 3n^2 + 3n^2 - 12 = 0$ Combine similar terms.
$2n - 12 = 0$ $\left\{\begin{array}{l}\text{To undo the subtraction of 12 from } 2n, \text{ add 12 to}\\ \text{each side. To undo the multiplication of } n \text{ by 2,}\\ \text{divide each side by 2.}\end{array}\right.$
$2n = 12$
$n = 6$

The solution set is $\{6\}$.

---

olve.

. $2(x - 1) + 3 = 7$

34. $3(y - 2) + 1 = 10$

. $2(2m - 3) - 3(2m - 1) = 9$

36. $4(3a - 1) - 5(1 - a) = 8$

. $y(3 - 2y) + 2(y^2 - 6) = 0$

38. $0 = 3(1 - 2x) - 5(3 - x)$

. $x(2 - 3x) + 3(x^2 - 6) = 0$

40. $2x(1 - 3x) + 6(x^2 - 2) = 0$

ixed Review Exercises

mplify.

. $(2xy^2)^3$

2. $(-4a^4b^3)^2$

. $(-2n)^4$

4. $(2a^2b)^2(3ab^2)^3$

. $(3x^2)(4x^3) + (2x^3)(5x^2)$

6. $(4n^3)n^2 - n^3(3n^2)$

. $(6p - 2q + 4) + (2p + 3q)$

8. $(3x + y - 2) - (y - x - 3)$

. $(4x^3)^2(2x^2y)^3$

10. $(7x^5)(2x) + (6x^4)(4x^2)$

NAME _____ DATE _____

4–6 Multiplying Polynomials

Objective: To multiply polynomials.

Example 1   Multiply: $(2x - 3)(x^2 - 4x - 5)$

Solution   You can find the product by arranging your work in vertical form. Each term of one polynomial must be multiplied by each term of the other polynomial.

Step 1:
Multiply by $2x$.

$$x^2 - 4x - 5$$
$$\underline{2x - 3}$$
$$2x^3 - 8x^2 - 10x$$

Step 2:
Multiply by -3.

$$x^2 - 4x - 5$$
$$\underline{2x - 3}$$
$$2x^3 - 8x^2 - 10x$$
$$\quad\quad\; - 3x^2 + 12x + 15$$
$$\quad\quad\; \uparrow \quad\quad \uparrow$$

Align similar terms.

Step 3:
Add the results of Steps 1 and 2.

$$x^2 - 4x - 5$$
$$\underline{2x - 3}$$
$$2x^3 - 8x^2 \quad - 10x$$
$$\underline{\quad\quad - 3x^2 \quad + 12x + 15}$$
$$2x^3 - 11x^2 + 2x \; + 15$$

Multiply. Use the vertical form.

1. $2a + 1$
    $\underline{\;a + 6}$

2. $3n + 6$
    $\underline{2n - 5}$

3. $3x - 7$
    $\underline{2x + 1}$

4. $4t - 1$
    $\underline{3t - 2}$

5. $3x - 4y$
    $\underline{5x - 2y}$

6. $2c - 5d$
    $\underline{3c + \;d}$

7. $5c - 3d$
    $\underline{2c + \;d}$

8. $3x^2 - x - 4$
    $\underline{x + 4}$

9. $a^2 - 5a - 7$
    $\underline{3a + 2}$

10. $4y^2 - 5y - 2$
    $\underline{2y - 1}$

11. $a^2 - ab + b^2$
    $\underline{a + b}$

12. $2x^2 - xy + y^2$
    $\underline{2x + y}$

Example 2   Multiply: $(3x - 2)(2x + 5)$

Solution $(3x - 2)(2x + 5) = (3x - 2)2x + (3x - 2)5$ Use the distributive property.
$$= 6x^2 - 4x + 15x - 10 \quad \text{Combine like terms.}$$
$$= 6x^2 + 11x - 10$$

Multiply. Use the horizontal form.

13. $(a + 2)(a + 3)$

14. $(b + 4)(b + 5)$

15. $(x - 3)(x + 8)$

16. $(c + 1)(c - 4)$

17. $(2a - 1)(a + 4)$

18. $(3a + 4)(a - 1)$

19. $(2a + 3)(5a - 1)$

20. $(4k - 5)(2k + 6)$

21. $(x - 1)(2x^2 + 3x + 4)$

22. $(2a + 1)(a^2 + 2a + 5)$

23. $(t - 3)(3t^2 + 3t - 4)$

24. $(t - 2)(2t^2 - 3t - 4)$

25. $(2x - 3)(3x^2 - 4x - 2)$

26. $(3x - 4)(2x^2 - x + 1)$

–6 *Multiplying Polynomials* (continued)

CAUTION It often is helpful to rearrange the terms of a polynomial so that the degrees of a particular variable are in either increasing order or decreasing order. For example:

In order of decreasing degree of x:

$$x^4 - 2x^3 - 5x + 6$$

In order of increasing degree of x:

$$6 - 5x - 2x^3 + x^4$$

In order of decreasing degree of x and increasing degree of y:

$$x^4 - 5x^3y + 3x^2y^2 - 6xy^3 + 9y^4$$

Example 3   Multiply: $(y + 3x)(x^3 - y^3 + 2x^2y + 3xy^2)$

Solution $x^3 - y^3 + 2x^2y + 3xy^2$
$y + 3x$
_____

Rearrange in order of decreasing degree of x and increasing degree of y.

$x^3 + 2x^2y + 3xy^2 - y^3$
$3x + y$
_____
$3x^4 + 6x^3y + 9x^2y^2 - 3xy^3$
$\quad\quad x^3y + 2x^2y^2 + 3xy^3 - y^4$
_____
$3x^4 + 7x^3y + 11x^2y^2 \quad\quad - y^4$

Therefore $(y + 3x)(x^3 - y^3 + 2x^2y + 3xy^2) = 3x^4 + 7x^3y + 11x^2y^2 - y^4$.

Multiply using either the horizontal or vertical form. Arrange the terms in each factor in order of decreasing or increasing degree of one of the variables.

. $(1 + y)(y^2 + 2y - 3)$ **28.** $(4 + x)(x^2 - 4x + 3)$

. $(2 + 3y)(3y - 5 + y^2)$ **30.** $(3y + 4)(y - 2y^2 + 5)$

. $(3x + y)(x^2 + 4y^2 + 2xy)$ **32.** $(1 + 2a)(a^2 - 4 + a)$

. $(2x - y)(x^2 + 3y^2 - 4xy)$ **34.** $(y - 3x)(2x^2 + y^2 - 2xy)$

Mixed Review Exercises

Solve.

. $2(x - 1) = 8$ **2.** $3(x - 2) - 2 = 7$ **3.** $4(2a + 3) = 5(a - 6)$

Evaluate if $w = -1$, $x = 2$, and $y = 4$.

. $x + |w| - y$ **5.** $w + x + y$ **6.** $w - |y - x|$

. $(x + y)^2$ **8.** $(-x)^2x^2$ **9.** wy^3

4–7 Transforming Formulas

Objective: To transform a formula.

Example 1 Solve the formula $F = ma$ for m. State the restrictions, if any, for the formula obtained to be meaningful.

Solution $F = ma$ To get m alone on one side, divide both sides by a.

$\dfrac{F}{a} = m, a \neq 0$ The denominator cannot be 0.

Solve the given formula for the indicated variable. State the restrictions, if any, for the formula obtained to be meaningful.

1. $C = \pi d$ for d

2. $F = ma$ for a

3. $I = prt$ for t

4. $V = Bh$ for h

5. $d = rt$ for t

6. $s = gt^2$ for g

Example 2 The formula $A = \dfrac{1}{2}h(a + b)$ gives the area of a trapezoid with bases a units and b units and with height h units. Use this formula to solve for the variable b in terms of A, h, and a. State the restrictions, if any, for the formula obtained to be meaningful.

Solution $A = \dfrac{1}{2}h(a + b)$ To get clear of fractions, multiply both sides by 2.

$2A = h(a + b)$ Divide both sides by h.

$\dfrac{2A}{h} = a + b$ Subtract a from both sides.

$\dfrac{2A}{h} - a = b, h \neq 0$ The denominator cannot be 0.

Solve the given formula for the indicated variable. State the restrictions, if any, for the formula obtained to be meaningful.

7. $A = \dfrac{1}{2}bh$ for h

8. $b = 2b + y$ for y

9. $A = \dfrac{1}{2}h(b + c)$ for h

10. $A = P + Prt$ for r

11. $a = 2(l + w)$ for l

12. $C = \dfrac{5}{9}(F - 32)$ for F

–7 Transforming Formulas (continued)

Example 3 Solve the formula $C = \dfrac{mv^2}{r}$ for r. State the restrictions, if any, for the formula obtained to be meaningful.

Solution $C = \dfrac{mv^2}{r}$ To get r out of the denominator, multiply both sides by r.

$Cr = mv^2$ To get r alone on one side, divide both sides by C.

$r = \dfrac{mv^2}{C}, \; C \neq 0$ The denominator cannot be 0.

Solve the given formula for the indicated variable. State the restrictions, if any, for the formula obtained to be meaningful.

] $s = \dfrac{v}{r}$ for v

14. $d = \dfrac{m}{v}$ for m

. $C = \dfrac{mv^2}{r}$ for m

16. $2ax + 1 = ax + 5$ for x

. $a = \dfrac{v - u}{t}$ for u

18. $v^2 = u^2 + 2as$ for a

. $S = \dfrac{n}{2}(a + 1)$ for a

20. $m = \dfrac{x + y + z}{3}$ for x

. $l = a + (n - 1)d$ for d

22. $A = \dfrac{a + b + c + d}{4}$ for b

. $3by - 2 = 2by + 1$ for b

24. $3aw + 1 = aw - 7$ for a

. $ax + b = c$ for b

26. $D = \dfrac{a}{2}(2t - 1)$ for a

. $am - bm = c$ for a

28. $q = 1 + \dfrac{P}{100}$ for P

ixed Review Exercises

implify.

. $(y - 4)(y + 2)$

2. $(2n - 3)(3n - 4)$

. $a[3a - 2(4 + a)]$

4. $xy(x - 2y)$

. $3x(x^2 - 2x + 3)$

6. $(-4x^2)^3$

. $n^2 \cdot n^3 \cdot n^4$

8. $(2a^2)^3 \cdot (3a^3b^2)$

. $(x + 6)(x - 5)$

10. $(a + 2b)ab$

. $(4m + 5)(8m + 7)$

12. $2y^2(y^3 + 2y - 1)$

NAME _____ DATE _____

4–8 Rate-Time-Distance Problems

Objective: To solve some word problems involving uniform motion.

Vocabulary   **Uniform motion** Motion at a constant speed, or rate.

CAUTION   Some problems give the time in *minutes* and the speed in *miles per hour*.
Be sure to write the time in terms of hours when you use the given facts.

Example 1   **(Motion in opposite directions)** Two jets leave St. Louis at 8:00 A.M., one flying east at a speed 40 km/h greater than the other, which is traveling west. At 10:00 A.M. the planes are 2480 km apart. Find their speeds.

Solution

Step 1   The problem asks for the speeds, or rates, of the planes.

Step 2   Let r = the rate of the plane flying west. Then $r + 40$ = rate of the plane flying east. Make a chart organizing the given facts and use it to label a sketch. Remember that east and west are *opposite directions*.

	Rate	× Time =	Distance
East bound	$r + 40$	2	$2(r + 40)$
West bound	r	2	$2r$

$2r \qquad$ St. Louis $\cdot 2(r + 40)$

2480 km

Step 3   The distance between two objects moving in *opposite directions* is the *sum* of the separate distances traveled. The sum of the distances is 2480 km.
$$2r + 2(r + 40) = 2480$$

Step 4
$$2r + 2r + 80 = 2480$$
$$4r + 80 = 2480$$
$$4r = 2400$$
$$r = 600$$
$$r + 40 = 640$$

Step 5   *Check:*   In 2 h the eastbound plane flies 2(640) = 1280 (km). In 2 h the westbound plane flies 2(600) = 1200 (km). 1280 + 1200 = 2480 √

The speed of the plane flying west is 600 km/h.
The speed of the plane flying east is 640 km/h.

Solve.

1. Two jets leave Ontario at the same time, one flying east at a speed 20 km/h greater than the other, which is flying west. After 4 h, the planes are 6000 km apart. Find their speeds.

2. Two camper vans leave Arrowhead Lake at the same time, one traveling north at a speed of 10 km/h faster than the other, which is traveling south. After 3 h, the camper vans are 420 km apart. Find their speeds.

3. Two cars traveled in opposite directions from the same starting point. The rate of one car was 10 km less than the rate of the other. After 4 h the cars were 600 km apart. Find the rate of each car.

Example 2   **(Motion in the same direction)** A small plane leaves an airport and flies north at 240 mi/h. A jet leaves the airport 30 min later and follows the small plane at 360 mi/h. How long does it take the jet to overtake the small plane?

Solution

Step 1   The problem asks for the jet's flying time before it overtakes the small plane.

Step 2   Let t = the jet's flying time. Then $t + \frac{1}{2}$ = the small plane's flying time. Make a chart organizing the given facts and use it to label a sketch. Notice that 30 min must be written as $\frac{1}{2}$ h.

	Rate $\times$ Time	= Distance
Plane	240	$t + \frac{1}{2}$	$240(t + \frac{1}{2})$
Jet	360	t	$360t$

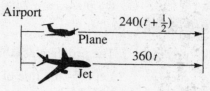

Step 3   When the jet overtakes the plane, the distances will be equal.
$$360t = 240(t + \tfrac{1}{2})$$

Step 4 $360t = 240t + 120$
$120t = 120$
$t = 1$

Step 5   Check in the words of the problem. The jet overtakes the plane in 1 h.

Solve.

4. A car started out from Memphis toward Little Rock at the rate of 60 km/h. A second car left from the same point 2 h later and drove along the same route at 75 km/h. How long did it take the second car to overtake the first car?

5. A tourist bus leaves Richmond at 1:00 P.M. for New York City. Exactly 24 min later, a truck sets out in the same direction. The tourist bus moves at a steady 60 km/h. The truck travels at 80 km/h. How long does it take the truck to overtake the tourist bus?

6. Exactly 20 min after Alex left home, his sister Alison set out to overtake him. Alex drove at 48 mph and Alison drove at 54 mph. How long did it take Alison to overtake Alex?

7. The McLeans drove from their house to Dayton at 75 km/h. When they returned, the traffic was heavier and they drove at 50 km/h. If it took them 1 h longer to return than to go, how long did it take them to drive home?

8. It takes a plane 1 h less to fly from San Diego to New Orleans at 600 km/h than it does to return at 450 km/h. How far apart are the cities?

Mixed Review Exercises

Solve.

1. $32 = -8x$

2. $3(x + 4) = 27$

3. $(x - 4)(x + 7) = (x + 4)(x - 3)$

NAME _____   DATE _____

4–9 Area Problems

Objective: To solve some problems involving area.

Formula

Area of rectangle = length × width

Example   A rectangular flower garden is 8 ft longer than it is wide. It is surrounded by a brick walk 2 ft wide. The area of the walk is 176 ft². Find the dimensions of the flower garden.

Solution

Step 1   The problem asks for the dimensions of the flower garden. Make a sketch.

Step 2   Let x = the width of the flower garden.
Then $x + 8$ = the length of the flower garden.
Label your sketch.

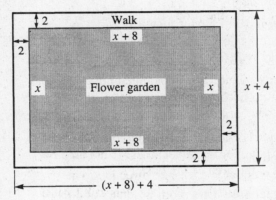

Step 3   Area of walk = Area of garden and walk − Area of garden
$$176 = (x + 12)(x + 4) - x(x + 8)$$

Step 4 $176 = x^2 + 16x + 48 - x^2 - 8x$
$176 = 8x + 48$
$128 = 8x$
$16 = x$ and $x + 8 = 24$

Step 5   *Check:* If the dimensions of the flower garden are 16 ft and 24 ft,
the dimensions of the flower garden and walk are 20 ft and 28 ft.

Area of the flower garden and walk = 20 · 28 = 560 (ft²)
Area of the flower garden = 16 · 24 = 384 (ft²)
Area of walk = 560 − 384 = 176 (ft²)

The dimensions of the flower garden are 24 ft and 16 ft.

4–9 *Area Problems* (continued)

Solve.

1. A rectangle is twice as long as it is wide. If its length and width are both decreased by 4 cm, its area is decreased by 164 cm². Find its original dimensions.

2. A rectangle is three times as long as it is wide. If its length and width are both increased by 3 m, its area is increased by 81 m². Find its original dimensions.

3. A rectangle is 8 cm longer than it is wide. If its length and width are both increased by 2 cm, its area is increased by 68 cm². Find its original dimensions.

4. A rectangle is 12 cm longer than it is wide. If its length and width are both decreased by 2 cm, its area is decreased by 108 cm². Find its original dimensions.

5. A rectangular fish pond is 8 ft longer than it is wide. A wooden walk 2 ft wide is placed around the pond. The area covered by the pond and walk is 160 ft² greater than the area covered by the pond alone. What are the dimensions of the pond?

6. A rectangular swimming pool is 10 m longer than it is wide. A walkway 2 m wide surrounds the pool. Find the dimensions of the pool if the area of the walkway is 216 m².

7. A poster is 24 cm taller than it is wide. If it is mounted on a piece of cardboard so that there is a 6 cm border on all sides and if the area of the border alone is 720 cm², what are the dimensions of the poster?

8. A poster is 20 cm longer than it is wide. It is mounted on a piece of cardboard so that there is a 10 cm border on all sides. If the area of the border is 2400 cm², what is the area of the poster?

9. A house has two rooms of equal area. One room is square and the other is 6 ft narrower and 9 ft longer than the square room. Find the area of each room.

10. A baker has two pans with the same area. One pan is square and the other is 6 cm narrower and 8 cm longer than the square pan. Find the area of each pan.

Mixed Review Exercises

Simplify.

1. $(-3 + 7) + (-6)$
2. $4(c - 1) + (2)c + 7$
3. $(3x^2y)^4$

4. $y - (-5) - [y + (-3)]$
5. $4^3 \cdot 2^2$
6. $(5 \cdot 4 + 3 \cdot 4) \div (4 \cdot 2)$

7. $-\frac{8}{3} + \frac{2}{3} + 2$
8. $\left(\frac{1}{3}x^3\right)\left(\frac{3}{5}x^2\right)$
9. $-\frac{x}{8}(-96)$

10. $(-7 + 11) + (-4)$
11. $\left(\frac{1}{2}x^2\right)\left(\frac{2}{5}x\right)$
12. $(2x^2y)^3$

4–10 Problems Without Solutions

Objective: To recognize problems that do not have solutions.

CAUTION Not all word problems have solutions. Here are some reasons for this:

1. Not enough information is given.
2. The given facts lead to an unrealistic result. (The result satisfies the equation used, but not the conditions of the problem situation.)
3. The given facts are contradictory. (They cannot all be true at the same time.)

Example 1 On a trip of 210 km, Roberto went by train for 3h and by bus for the rest of the trip. The average speed of the train was 15 km/h more than that of the bus. Find the average speed of the bus.

Solution

Step 1 The problem asks for the speed of the bus.

Step 2 Let r = the speed of the bus.
Then $r + 15$ = the speed of the train.

Make a chart showing the given facts.

	Rate	× Time =	Distance
Train trip	$r + 15$	3	$3(r + 15)$
Bus trip	r	?	?

All of the given facts have been used, but *not enough information has been given to write an equation.*

The problem does not have a solution.

Example 2 Angela says she has the same number of nickels as dimes. Three times the value of the dimes is equal to 20¢ more than six times the value of the nickels. How many of each kind does she have?

Solution

Step 1 The problem asks for the number of nickels and dimes.

Step 2 Let x = the number of nickels and let x = the number of dimes.
Then $5x$ = the value (¢) of the nickels and $10x$ = the value (¢) of the dimes.

Step 3 $3(10x) = 20 + 6(5x)$

Step 4 $30x = 20 + 30x$
$0 = 20$

The false statement "$0 = 20$" tells you the *given facts are contradictory.*

The problem has no solution.

NAME _____ DATE _____

Example 3   A bank teller cashes a check for $100 and gives the customer the same number of $10 bills and $20 bills. How many of each type of bill will the customer get?

Solution

Step 1   The problem asks for the number of bills of each type the customer will get.

Step 2   Let x = the number of each kind of bill.

Step 3 $10x + 20x = 100$

Step 4 $30x = 100$ $\left\{ \begin{array}{l} \text{Since you can't have } \frac{1}{3} \text{ of a bill, the given facts lead to an} \\ \text{\textit{unrealistic result}. The problem does not have a solution.} \end{array} \right.$
$x = 3\frac{1}{3}$

Solve each problem that has a solution. If a problem has no solution, explain why.

1. You plan a lawn 10 ft longer than it is wide. It is to be surrounded by a flower bed 3 ft wide. Find the dimensions of the lawn if the flower bed covers 80 ft².

2. A bank teller was asked to cash a check for $180 and to give the customer the same number of $10 bills and $20 bills. How many bills of each kind did the teller count out?

3. In the course of a year, the sum of an investor's gains and losses was $2400. What were his gains that year?

4. Find three consecutive integers whose sum is 47 more than the largest integer.

5. Find two consecutive integers whose sum is 48.

6. Chan bought more 25¢ stamps than 15¢ stamps. How many of each kind did he buy if the total cost of the stamps is $7.80?

7. Hans has as many dimes as quarters. The dimes and quarters total $10. How many of each kind does he have?

8. Last month Charlene worked as many 8 h shifts as she worked 10 h shifts. How many 8 h shifts did she work last month?

9. Georgette has the same number of 25¢ stamps as she does 15¢ stamps. Six times the value of the 25¢ stamps is a dollar more than 10 times the value of the 15¢ stamps. How many of each type of stamp does she have?

10. Last month Theo worked the same number of hours on 8 h shifts as he did on 6 h shifts. One fourth the amount of time he put in at 8 h shifts was 1 h less than one third the time he put in at 6 h shifts. How many hours did he work at each type of shift?

Mixed Review Exercises

Solve.

1. $-5x = 40$

2. $2(n - 3) = 28$

3. $32 = -4x$

4. $-n + 6 = 2$

5. $x - 5 = |3 - 7|$

6. $-y + 15 = 10$

7. $-\frac{1}{2}(x + 4) = 3$

8. $\frac{1}{5}x = 12$

5 Factoring Polynomials

5–1 Factoring Integers

Objective: To factor integers and to find the greatest common factor of several integers.

Vocabulary

Factor To write a number as a product of numbers. For example, $72 = 8 \cdot 9$.

Factor set The set over which a number is factored.

Prime number, or prime An integer greater than 1 that has no positive integral factor other than itself and 1. For example, 19 is prime.

Prime factorization Writing a positive integer as a product of primes. For example, the prime factorization of 30 is $2 \cdot 3 \cdot 5$.

Common factor A factor of two or more integers is called a common factor of the integers. For example, 3 is a common factor of 6 and 9.

Greatest common factor (GCF) The greatest integer that is a factor of two or more given integers.

Example 1　List all the positive factors of 42.

Solution　
$$42 = 1 \cdot 42$$
$$= 2 \cdot 21$$
$$= 3 \cdot 14$$
$$= 6 \cdot 7$$
$$(= 7 \cdot 6)$$

{ Divide 42 by 1, 2, 3, ... until a pair of factors is repeated.

The positive factors of 42 are 1, 2, 3, 6, 7, 14, 21, and 42.

List all of the positive factors of each number.

1. 10　　　**2.** 24　　　**3.** 36　　　**4.** 40

5. 17　　　**6.** 54　　　**7.** 29　　　**8.** 42

CAUTION　Factors come in pairs. For example, since $12 \div 3 = 4$, 3 and 4 are both factors of 12.

Example 2　List all pairs of factors of each integer:　**a.** 18　**b.** −18

Solution　
a. (1)(18)　(−1)(−18)　　**b.** (1)(−18)　(−1)(18)
　　(2)(9)　(−2)(−9)　　　　(2)(−9)　(−2)(9)
　　(3)(6)　(−3)(−6)　　　　(3)(−6)　(−3)(6)

List all pairs of factors of each integer.

9. 11　　　**10.** 20　　　**11.** 23　　　**12.** 39　　　**13.** 57

14. 60　　　**15.** 75　　　**16.** 78　　　**17.** 81　　　**18.** 105

19. 121　　　**20.** −30　　　**21.** −63　　　**22.** −57　　　**23.** −93

–1 Factoring Integers (continued)

Example 3 Find the prime factorization of 252.

Solution Try the primes in order as divisors. Divide by each prime as many times as possible before going on to the next prime. Stop when all factors are primes.

$$252 = 2 \cdot 126$$
$$= 2 \cdot 2 \cdot 63$$
$$= 2 \cdot 2 \cdot 3 \cdot 21$$
$$= 2 \cdot 2 \cdot 3 \cdot 3 \cdot 7$$
$$= 2^2 \cdot 3^2 \cdot 7$$

Find the prime factorization of each number. A calculator may be helpful.

4. 22	**25.** 30	**26.** 56	**27.** 64	**28.** 44	**29.** 50
). 72	**31.** 84	**32.** 93	**33.** 180	**34.** 275	**35.** 388

Example 4 Find the GCF of 540 and 264.

Solution
1. First find the prime factorization of each integer.
$$540 = 2^2 \cdot 3^3 \cdot 5 \qquad 264 = 2^3 \cdot 3 \cdot 11$$
2. Then find the product of smaller powers of each common prime factor.

   The common prime factors are 2 and 3.
   The smaller power of 2 is 2^2.
   The smaller power of 3 is 3.
3. The GCF of 540 and 264 is $2^2 \cdot 3$ or 12.

CAUTION If there are no common prime factors, the GCF is 1. For example, since $12 = 2^2 \cdot 3$ and $25 = 5^2$, the GCF of 12 and 25 is 1.

Find the GCF of each group of numbers. A calculator may be helpful.

. 36, 90	**37.** 28, 70	**38.** 120, 128	**39.** 108, 180
. 105, 350	**41.** 126, 144	**42.** 145, 174	**43.** 260, 325

ixed Review Exercises

mplify.

. $\frac{1}{2}(4x + 2) + 3\left(\frac{1}{3}x - 1\right)$ **2.** $(4 + 3)^2$ **3.** $2^2 + (3 + 1)^2$

. $2x - 3 - (2x + 4)$ **5.** $2ab(3a^2)(4b)$ **6.** $2x^3(3y)(5y)$

. $(2x)^3$ **8.** $3n(2n^2 - 5n) + 7n^2$ **9.** $(-3)^4x^4$

. $x(x^2 - 2) - x^2(x + 4)$ **11.** $(3y + 4)(y + 2)$ **12.** $(x - 3)(2x + 3)$

5–2 *Dividing Monomials*

Objective: To simplify quotients of monomials and to find the GCF of several monomials.

Vocabulary

Greatest common factor (GCF) of two or more monomials The common
factor with the *greatest coefficient* and the *greatest degree* in each
variable. For example, $5x^2y$ is the GCF of $10x^2y^2$ and $25x^3y$.

Properties and Rules

Property of Quotients

If a, b, c, and d are real numbers with $b \neq 0$, and $d \neq 0$, then $\dfrac{ac}{bd} = \dfrac{a}{b} \cdot \dfrac{c}{d}$.

For example, $\dfrac{15}{16} = \dfrac{3 \cdot 5}{2 \cdot 8} = \dfrac{3}{2} \cdot \dfrac{5}{8}$

Rule for Simplifying Fractions

If b, c, and d are real numbers with $b \neq 0$ and $d \neq 0$, then $\dfrac{bc}{bd} = \dfrac{c}{d}$.

For example, $\dfrac{15}{18} = \dfrac{3 \cdot 5}{3 \cdot 6} = \dfrac{5}{6}$.

Rule of Exponents for Division

If a is a nonzero real number and m and n are positive integers, then:

If $m > n$:   If $n > m$:   If $m = n$:

$\dfrac{a^m}{a^n} = a^{m-n}$ $\dfrac{a^m}{a^n} = \dfrac{1}{a^{n-m}}$ $\dfrac{a^m}{a^n} = 1$

CAUTION   You can divide the numerator and denominator of a fraction only by
a nonzero number. In the examples of this lesson, *assume that no
denominator equals zero.*

Example 1   Simplify: **a.** $\dfrac{28}{35}$ **b.** $\dfrac{-15xy}{21x}$

Solution   **a.** Divide both numerator and denominator by 7. The "cancel" marks show this.

$$\frac{28}{35} = \frac{4 \cdot \cancel{7}}{5 \cdot \cancel{7}} = \frac{4}{5}$$

b. Divide both numerator and denominator by $3x$.

$$\frac{-15xy}{21x} = \frac{\cancel{3x}(-5y)}{\cancel{3x} \cdot 7} = \frac{-5y}{7}, \text{ or } -\frac{5y}{7}$$

Example 2   Simplify: **a.** $\dfrac{x^8}{x^3}$ **b.** $\dfrac{x^3}{x^8}$ **c.** $\dfrac{x^2}{x^2}$

Solution   **a.** $\dfrac{x^8}{x^3} = x^{8-3} = x^5$ **b.** $\dfrac{x^3}{x^8} = \dfrac{1}{x^{8-3}} = \dfrac{1}{x^5}$ **c.** $\dfrac{x^2}{x^2} = 1$

5–2 *Dividing Monomials* (continued)

Simplify. Assume that no denominator equals zero.

1. $\dfrac{25}{30}$ **2.** $\dfrac{48}{72}$ **3.** $\dfrac{54}{72}$ **4.** $\dfrac{10^3}{10^6}$ **5.** $\dfrac{10^8}{10^5}$ **6.** $\dfrac{10a}{2a}$

7. $\dfrac{12m}{4m}$ **8.** $\dfrac{15 \cdot 10^3}{5 \cdot 10^4}$ **9.** $\dfrac{6x^4}{9x^2}$ **10.** $\dfrac{4n^6}{20n^4}$ **11.** $\dfrac{2x^5}{16x^4}$ **12.** $\dfrac{12y^3}{3xy^2}$

3. $\dfrac{4a^2b}{16ab^2}$ **14.** $\dfrac{-6x^2y^3}{9xy^2}$ **15.** $\dfrac{-8a^2b}{-20ab}$ **16.** $\dfrac{-32cd^3}{-24bd^2}$ **17.** $\dfrac{-21bc^3}{-14cd^2}$

8. $\dfrac{30xz^3}{-35yz^2}$ **19.** $\dfrac{x^2yz^3}{x^3y^3z^3}$ **20.** $\dfrac{a^2b^4c}{a^2bc^3}$ **21.** $\dfrac{35a^2b^3c}{25abc}$ **22.** $\dfrac{26x^2yz}{52xyz}$

Example 3 $\dfrac{(9x)^2}{(3x)^3} = \dfrac{81x^2}{27x^3} = \dfrac{\cancel{27x^2} \cdot 3}{\cancel{27x^2} \cdot x} = \dfrac{3}{x}$

Simplify. Assume that no denominator equals zero.

3. $\dfrac{(2x)^3}{2x^3}$ **24.** $\dfrac{5m^2}{(5m)^2}$ **25.** $\dfrac{(2t^2)^3}{(2t^3)^2}$ **26.** $\dfrac{(4a^2)^3}{(4a^3)^2}$ **27.** $\dfrac{(3ab)^2}{3a^2b}$

8. $\dfrac{(2mn)^3}{2m^3n^2}$ **29.** $\dfrac{(-z)^6}{(-z)^3}$ **30.** $\dfrac{(-a)^5}{(-a)^3}$ **31.** $\dfrac{(-xy)^7}{xy^7}$ **32.** $\dfrac{(-t^3)^4}{(-t^2)^5}$

Example 4   Find the missing factor. $45x^2y^3z^4 = (3xyz^2)(?)$ **Solution** $\dfrac{45x^2y^3z^4}{3xyz^2} = 15xy^2z^2$

Find the missing factor.

3. $8t^4 = (2t)(?)$ **34.** $10w^4 = (2w^2)(?)$ **35.** $6a^3b^5 = (2a^2b^2)(?)$

5. $15pq^3 = (5pq)(?)$ **37.** $-28x^2y^4 = (7x^2y)(?)$ **38.** $-32a^5b^4 = (-8a)(?)$

Example 5   Find the GCF of $18x^3y$ and $10x^2y^3$.

Solution
$$\left.\begin{array}{l} 18 = 2 \cdot 3^2 \\ 10 = 2 \cdot 5 \end{array}\right\} \text{GCF} = 2$$
$$\left.\begin{array}{l} x^3y \\ x^2y^3 \end{array}\right\} \text{GCF} = x^2y$$
$$\left.\vphantom{\begin{array}{l}1\\2\\3\\4\end{array}}\right\} \text{GCF} = 2x^2y$$

Find the GCF.

39. $21x^3$, $14x^2$
40. a^3b^2, a^2b^3
41. $6xy^2$, $8x^4y^3$
42. $18c^2d^3$, $24c^2d$
43. $35pq^2r$, $25p^3qr^2$

Mixed Review Exercises

Simplify.

. $\dfrac{1}{4}(-24)$ **2.** $105 \cdot \dfrac{1}{5}$ **3.** $378 \div 9$ **4.** $4n^3\left(\dfrac{1}{4}n^3\right)$ **5.** $12 \div \left(-\dfrac{1}{3}\right)$ **6.** $10y \cdot \dfrac{2}{5}y^2$

5–3 *Monomial Factors of Polynomials*

Objective: To divide polynomials by monomials and to find monomial factors of polynomials.

Vocabulary

Divisible One polynomial is (evenly) divisible by another polynomial if the quotient is also a polynomial. Example 1b shows that $27uv - 36v$ is divisible by $9v$.

Greatest monomial factor of a polynomial The GCF of the terms of the polynomial. In Example 3, the GCF of $3x^2 + 12x$ is $3x$.

Example 1 Divide: **a.** $\dfrac{6m + 36}{6}$ **b.** $\dfrac{27uv - 36v}{9v}$

Solution Divide each term of the polynomial by the monomial. Then add the results.

a. $\dfrac{6m + 36}{6} = \dfrac{6m}{6} + \dfrac{36}{6}$ **b.** $\dfrac{27uv - 36v}{9v} = \dfrac{27uv}{9v} - \dfrac{36v}{9v}$

$\qquad\qquad = m + 6$ $\qquad\qquad\qquad = 3u - 4$

Example 2 Divide: $\dfrac{2x^4 - 8x^3y + 4x^2y^2}{-2x^2}$

Solution $\dfrac{2x^4 - 8x^3y + 4x^2y^2}{-2x^2} = \dfrac{2x^4}{-2x^2} - \dfrac{8x^3y}{-2x^2} + \dfrac{4x^2y^2}{-2x^2} = -x^2 + 4xy - 2y^2$

Divide. Assume that no denominator equals zero.

1. $\dfrac{4a + 12}{4}$ 2. $\dfrac{10a - 15}{5}$

3. $\dfrac{20n - 16}{4}$ 4. $\dfrac{6x + 9y + 12}{3}$

5. $\dfrac{2m - 4n + 6}{2}$ 6. $\dfrac{x^3 - 4x^2 + 6x}{x}$

7. $\dfrac{8xy - 12x^2}{4x}$ 8. $\dfrac{5a - 10a^2 - 15a^3}{5a}$

9. $\dfrac{12y - 18y^2 - 6y^3}{6y}$ 10. $\dfrac{4x^2 - 12x - 8}{4}$

11. $\dfrac{27y^4 + 18y^3 - 36y^2}{9y^2}$ 12. $\dfrac{6u^3 + 4u^2 - 2u}{2u}$

13. $\dfrac{12r^4 - 9r^3 - 6r^2}{-3r^2}$ 14. $\dfrac{5m^3 + 8m^4 - 3m^5}{-m^3}$

15. $\dfrac{xy^3 + x^3y}{xy}$ 16. $\dfrac{8ab^2 - 12a^2b}{4ab}$

5-3 Monomial Factors of Polynomials (continued)

Example 3 Factor $3x^2 + 12x$

Solution 1. The greatest monomial factor of $3x^2 + 12x$ is $3x$.

2. Divide to find the other factor: $\dfrac{3x^2 + 12x}{3x} = \dfrac{3x^2}{3x} + \dfrac{12x}{3x}$

$$= x + 4$$

3. $3x^2 + 12x = 3x(x + 4)$

Example 4 Factor $6x^5 - 4x^3 + 8x$

Solution 1. The greatest monomial factor of $6x^5 - 4x^3 + 8x$ is $2x$.

2. Divide to find the other factor: $\dfrac{6x^5 - 4x^3 + 8x}{2x} = \dfrac{6x^5}{2x} - \dfrac{4x^3}{2x} + \dfrac{8x}{2x}$

$$= 3x^4 - 2x^2 + 4$$

3. $6x^5 - 4x^3 + 8x = 2x(3x^4 - 2x^2 + 4)$

Factor.

17. $21a^3 - 14a^2$

18. $4x^3 + 32x$

19. $9x^2 + 36x$

20. $21c^3 - 14c$

21. $10a - 35b + 15$

22. $16x - 12y + 24$

23. $8p - 4q + 12$

24. $3x - 6y + 12$

25. $9x - 6y + 36$

26. $15a - 20b + 10$

27. $3a^3 + 6a^2 - 12a$

28. $10x^3 - 5x^2 + 20x$

29. $5y^3 - 10y^2 + 15y$

30. $18x^3 - 6x^2 + 24x$

31. $8ab^2 - 12a^2b$

32. $3a^2b^2 + 18ab$

33. $6y^3 - 24y^2 - 12y$

34. $20y^4 + 35y^3 + 15y^2$

Mixed Review Exercises

Simplify.

1. $6n^2\left(\dfrac{1}{6}n^2\right)$

2. $8x^2\left(\dfrac{3}{4}x^3\right)$

3. $3a^2 - 6ac^2 + 4a^2 - 5ac^2$

4. $\dfrac{5x^3y}{10x^2y^2}$

5. $24 \div \left(-\dfrac{1}{3}\right)$

6. $\dfrac{(3a^2)^3}{a^4}$

7. $(3a)^4$

8. $6(3^2 - 1) + 2^3$

9. $(x - 1)(x^2 + 2x + 3)$

10. $(m - 3)(m + 4)$

11. $(3a + 2)(5a - 3)$

12. $(6p - q)(2p - 3q)$

5–4 *Multiplying Binomials Mentally*

Objective: To find the product of two binomials mentally.

Vocabulary

Quadratic Term A term of degree two. For example, $2x^2$.

Linear Term A term of degree one. For example, $5x$.

Quadratic Polynomial A polynomial whose term of greatest degree is quadratic.
For example, $2x^2 - 5x + 7$.

Example 1   Write $(3x + 1)(4x - 5)$ as a trinomial.

Solution 1   You can work horizontally as shown at the left or vertically as shown at the right.

$$(3x + 1)(4x - 5) = 3x(4x - 5) + 1(4x - 5)$$
$$= 12x^2 - 15x + 4x - 5$$
$$= 12x^2 - 11x - 5$$

$$
\begin{array}{r}
4x - 5 \\
3x + 1 \\
\hline
12x^2 - 15x \\
4x - 5 \\
\hline
12x^2 - 11x - 5
\end{array}
$$

Solution 2   Use the FOIL method to multiply in your head.

Think of the products of these terms:

First Terms   Last Terms

$$(3x + 1)(4x - 5)$$

Inner
Terms

Outer Terms

Then write the products:

$$12x^2 - 15x + 4x - 5 = 12x^2 - 11x - 5$$

First   Outer   Inner   Last
terms   terms   terms   terms

Write each product as a trinomial.

1. $(x + 6)(x + 1)$

2. $(y + 3)(y + 4)$

3. $(a - 4)(a - 2)$

4. $(x - 5)(x - 6)$

5. $(c + 2)(c + 6)$

6. $(k - 3)(k - 6)$

7. $(a - 3)(a - 7)$

8. $(2 + x)(3 + x)$

9. $(k - 4)(k - 7)$

10. $(b - 2)(b + 7)$

11. $(c - 6)(c + 7)$

12. $(a - 4)(a - 6)$

13. $(2a + 3)(a + 4)$

14. $(3x + 2)(x + 4)$

5–4 *Multiplying Binomials Mentally* (continued)

Write each product as a trinomial.

15. $(2a + 7)(a - 2)$ **16.** $(4a - 1)(3a - 1)$

17. $(3a - 5)(2a - 1)$ **18.** $(3 - 2a)(2 - 3a)$

19. $(2k + 1)(3k + 4)$ **20.** $(3x - 2)(x + 5)$

21. $(4x + 3)(2x - 1)$ **22.** $(7m - 3)(6m + 2)$

Example 2 Write $(3x - 4y)(5x + y)$ as a trinomial.

Solution

$$\begin{aligned}
(3x - 4y)(5x + y) &= 15x^2 + (3xy - 20xy) - 4y^2 \\
&= 15x^2 - 17xy - 4y^2
\end{aligned}$$

with F, O, I, L labels

Write each product as a trinomial.

23. $(a - 2b)(a + b)$ **24.** $(x + 3y)(x + 2y)$

25. $(2x + y)(3x - 2y)$ **26.** $(3x + y)(x + 2y)$

27. $(4x + y)(2x - 3y)$ **28.** $(6a - b)(5a - 2b)$

Example 3 Write $(m^2 - 3m)(2m^2 + 5m)$ as a trinomial.

Solution

$$\begin{aligned}
(m^2 - 3m)(2m^2 + 5m) &= m^2(2m^2) + m^2(5m) + (-3m)(2m^2) + (-3m)(5m) \\
&= 2m^4 + (5m^3 - 6m^3) - 15m^2 \\
&= 2m^4 - m^3 - 15m^2
\end{aligned}$$

with F, O, I, L labels

Write each product as a trinomial.

29. $(x^2 - 2x)(2x^2 + 3x)$ **30.** $(a^2 - 3b)(2a^2 + b)$

31. $(u^2 + v^2)(u^2 - 4v^2)$ **32.** $(a^3 - 3b^3)(a^3 + 4b^3)$

33. $(x^3 - 2x)(x^3 + 1)$ **34.** $(x^3 - y^2)(3x^3 + y^2)$

Mixed Review Exercises

Simplify.

1. $(3x^2y)(-5x^2y^3)$ **2.** $(6x^2y^4)^3$ **3.** $(2n + 3)(3n^2 + n - 2)$

4. $\dfrac{15r^2 + 20r - 25}{5}$ **5.** $\dfrac{(4y)^3}{4y}$ **6.** $\dfrac{12 - 6x - 2x^2}{2}$

Solve.

7. $n = 32 - 3n$ **8.** $3x - (2x + 7) = 7$ **9.** $4(n + 1) = 3(4 + n)$

10. $5y + 3 = 53$ **11.** $2(x - 1) - 5 = 9$ **12.** $5(y - 2) + 4 = 14$

5–5 Differences of Two Squares

Objective: To simplify products of the form $(a + b)(a - b)$ and to factor differences of two squares.

Vocabulary

Product of the Sum and Difference of Two Numbers
$(a + b)(a - b) = a^2 - ab + ab - b^2 = a^2 - b^2$

Difference of Two Squares
$a^2 - b^2 = (a + b)(a - b)$

Example 1   Write each product as a binomial.

   a. $(x + 2)(x - 2)$ **b.** $(2n + 3)(2n - 3)$

Solution   These products fit the form $(a + b)(a - b)$, so each binomial is of the form $a^2 - b^2$.

   a. $(x + 2)(x - 2) = (x)^2 - (2)^2$
   $= x^2 - 4$

   b. $(2n + 3)(2n - 3) = (2n)^2 - (3)^2$
   $= 4n^2 - 9$

Write each product as a binomial.

1. $(a + 3)(a - 3)$ **2.** $(4 - x)(4 + x)$

3. $(x + 5)(x - 5)$ **4.** $(9 - x)(9 + x)$

5. $(5a + 2)(5a - 2)$ **6.** $(7a - 2)(7a + 2)$

7. $(4 + 3x)(4 - 3x)$ **8.** $(6 - 5x)(6 + 5x)$

9. $(3 - 5x)(3 + 5x)$ **10.** $(8x + 7)(8x - 7)$

Example 2   Write each product as a binomial.

   a. $(a^2 - 3b)(a^2 + 3b)$ **b.** $(xy + z)(xy - z)$

Solution   These products fit the form $(a + b)(a - b)$, so each binomial is of the form $a^2 - b^2$.

   a. $(a^2 - 3b)(a^2 + 3b) = (a^2)^2 - (3b)^2$
   $= a^4 - 9b^2$

   b. $(xy + z)(xy - z) = (xy)^2 - z^2$
   $= x^2y^2 - z^2$

Write each product as a binomial.

11. $(3x + 4y)(3x - 4y)$ **12.** $(2u + v)(2u - v)$ **13.** $(x^2 - 8y)(x^2 + 8y)$

14. $(x^2 - 3y^2)(x^2 + 3y^2)$ **15.** $(2a^2 + 5b^2)(2a^2 - 5b^2)$ **16.** $(ab - 2c)(ab + 2c)$

5–5 Differences of Two Squares (continued)

Example 3   Multiply. Use the pattern $(a + b)(a - b) = a^2 - b^2$.

   a. $58 \cdot 62$ **b.** $93 \cdot 87$

Solution   **a.** $58 \cdot 62 = (60 - 2)(60 + 2)$ **b.** $93 \cdot 87 = (90 + 3)(90 - 3)$
$$= 3600 - 4 \qquad\qquad\qquad\quad = 8100 - 9$$
$$= 3596 \qquad\qquad\qquad\qquad\quad = 8091$$

Multiply. Use the pattern $(a + b)(a - b) = a^2 - b^2$.

17. $16 \cdot 24$ **18.** $27 \cdot 33$ **19.** $53 \cdot 47$ **20.** $35 \cdot 45$

21. $41 \cdot 39$ **22.** $92 \cdot 88$ **23.** $104 \cdot 96$ **24.** $60 \cdot 140$

Example 4   Factor: **a.** $a^2 - 16$ **b.** $9 - 4b^2$ **c.** $25a^2 - 36x^6$

Solution   Use the pattern $a^2 - b^2 = (a + b)(a - b)$

   a. $a^2 - 16 = a^2 - 4^2$
$$= (a + 4)(a - 4)$$

   b. $9 - 4b^2 = 3^2 - (2b)^2$
$$= (3 + 2b)(3 - 2b)$$

   c. $25a^2 - 36x^6 = (5a)^2 - (6x^3)^2$
$$= (5a + 6x^3)(5a - 6x^3)$$

Factor. You may use a calculator or a table of squares.

25. $b^2 - 16$ **26.** $f^2 - 81$ **27.** $36 - x^2$

28. $9e^2 - 16$ **29.** $49n^2 - 1$ **30.** $4a^2 - 9$

31. $a^4 - 36$ **32.** $49b^2 - 16c^2$ **33.** $100 - 121r^2$

34. $121 - y^2$ **35.** $25u^2 - 36$ **36.** $16x^2 - 225$

Mixed Review Exercises

Simplify.

1. $5z(z - 2) + 3z(z + 4)$ **2.** $(x + 4)(x - 5)$ **3.** $-3(m + 2) - 4m(m - 3)$

4. $\dfrac{36a^5b^2}{9a^3}$ **5.** $\dfrac{15a + 5}{5}$ **6.** $\dfrac{18n^2x}{6nx}$

7. $(a + 2)(2a - 1)$ **8.** $(3b + 2)(b - 1)$ **9.** $(4x)^2\left(\dfrac{1}{4}\right)^2 x$

10. $\dfrac{12y^3 + 28y^2 - 8y}{4y}$ **11.** $\dfrac{30x^3 + 45x^2 - 15x}{15x}$ **12.** $\dfrac{24x^3y^4z}{3x^3y^3z}$

5-6 Squares of Binomials

Objective: To find squares of binomials and to factor perfect square trinomials.

Vocabulary

Square of a Binomial

$$(a + b)^2 = a^2 + ab + ab + b^2 = a^2 + 2ab + b^2$$
$$(a - b)^2 = a^2 - ab - ab + b^2 = a^2 - 2ab + b^2$$

Perfect Square Trinomial

$$a^2 + 2ab + b^2 = (a + b)^2$$
$$a^2 - 2ab + b^2 = (a - b)^2$$

Example 1   Write each square as a trinomial.

a. $(x + 4)^2$ b. $(5u - 2)^2$ c. $(2x + 3y)^2$ d. $(5a^2 - 4b^2)^2$

Solution   Use the patterns for the square of a binomial.

a. $(x + 4)^2 = x^2 + 2(x \cdot 4) + 4^2$ Pattern: $(a + b)^2 = a^2 + 2ab + b^2$
$\qquad\quad = x^2 + 8x + 16$

b. $(5u - 2)^2 = (5u)^2 - 2(5u \cdot 2) + 2^2$ Pattern: $(a - b)^2 = a^2 - 2ab + b^2$
$\qquad\qquad\; = 25u^2 - 20u + 4$

c. $(2x + 3y)^2 = (2x)^2 + 2(2x \cdot 3y) + (3y)^2$ Pattern: $(a + b)^2 = a^2 + 2ab + b^2$
$\qquad\qquad\quad = 4x^2 + 12xy + 9y^2$

d. $(5a^2 - 4b^2)^2 = (5a^2)^2 - 2(5a^2 \cdot 4b^2) + (4b^2)^2$ Pattern:
$\qquad\qquad\qquad = 25a^4 - 40a^2b^2 + 16b^4$ $(a - b)^2 = a^2 - 2ab + b^2$

Write each square as a trinomial.

1. $(x + 3)^2$

2. $(y - 2)^2$

3. $(a - 4)^2$

4. $(n + 5)^2$

5. $(3u - 1)^2$

6. $(5c - 1)^2$

7. $(3y + 4)^2$

8. $(4k + 5)^2$

9. $(5k - 2)^2$

10. $(ab - 3)^2$

11. $(3p + 5q)^2$

12. $(4x - 3y)^2$

13. $(3y - 5)^2$

14. $(5a - 7b)^2$

15. $(2x + y)^2$

16. $(x^2 + 5)^2$

17. $(ef - 7)^2$

18. $(pq - 3)^2$

19. $(2ab - c)^2$

20. $(-3ab + b^2)^2$

21. $(-4 + 3f)^2$

22. $(-11u + v^2)^2$

23. $(5p^3 - 6)^2$

24. $(-9t^2 - 2)^2$

–6 Squares of Binomials (continued)

Example 2 Decide whether each trinomial is a perfect square. If it is, factor it.

a. $4x^2 - 12x + 9$ **b.** $16u^2 + 20uv + 25v^2$

Solution   **a.** $4x^2 - 12x + 9$

1. Is the first term a square?   Yes; $4x^2 = (2x)^2$
2. Is the last term a square?   Yes; $9 = (3)^2$
3. Is the middle term, neglecting the sign,   Yes; $12x = 2(2x \cdot 3)$
   twice the product of $2x$ and 3?

Since the answers to Questions 1–3 are all Yes,
$4x^2 - 12x + 9$ is a perfect square.

Use the pattern $a^2 - 2ab + b^2 = (a - b)^2$:
$$4x^2 - 12x + 9 = (2x - 3)^2.$$

b. $16u^2 + 20uv + 25v^2$

1. Is the first term a square?   Yes; $16u^2 = (4u)^2$
2. Is the last term a square?   Yes; $25v^2 = (5v)^2$
3. Is the middle term, neglecting the sign,   No; $20uv \neq 2(4u \cdot 5v)$
   twice the product of $4u$ and $5v$?

Since the answer to Question 3 is No,
$16u^2 + 20uv + 25v^2$ is not a perfect square.

**ecide whether each trinomial is a perfect square. If it is, factor it.
 it is not, write *not a perfect square*.**

5. $n^2 - 4n + 4$ 26. $k^2 + 10k + 25$ 27. $a^2 - 6a + 9$

3. $y^2 - 8y + 16$ 29. $a^2 - 4a + 16$ 30. $81 + 18y + y^2$

1. $9x^2 - 12x + 4$ 32. $16k^2 - 40k + 25$ 33. $9y^2 + 48y + 64$

4. $9 + 6y + 4y^2$ 35. $25x^2 + 80xy + 64y^2$ 36. $4c^2 - 12c + 9$

7. $9n^2 - 24n + 16$ 38. $81 - 36k + 4k^2$ 39. $81k^2 + 180k + 100$

). $49a^2 - 42ab + 9b^2$ 41. $4m^2 - 36mn + 81n^2$ 42. $16x^2 - 24xy + 9y^2$

lixed Review Exercises

valuate if $x = 3$ and $y = 2$.

1. $x + y + (-6)$ 2. $x - |5 - y|$ 3. $6 + x^2y$

4. $(3 + xy)^2$ 5. $(x)^3(-y)^3$ 6. $(x^2y^2)^2$

mplify.

7. $(2x + 5)(2x - 5)$ 8. $(x - 6)(x + 2)$ 9. $(6 - 2)^3$

). $4 - 2^3$ 11. $\dfrac{(a^4)^3}{(a^2)^4}$ 12. $\dfrac{(3xy)^2}{9xy}$

5–7 *Factoring Pattern for x² + bx + c, c positive*

Objective: To factor quadratic trinomials whose quadratic coefficient is 1 and whose constant term is positive.

Vocabulary/Patterns

Factoring patterns for $x^2 + bx + c$ **when c is positive:**
When b is positive: $(x + ?)(x + ?)$.
When b is negative: $(x - ?)(x - ?)$.

Prime polynomial A polynomial with integral coefficients whose greatest
monomial factor is 1 and which can't be written as a product of
polynomials of lower degree. For example, $a^2 - 10a - 14$ is prime.

Example 1   Factor $x^2 + 6x + 8$.

Solution
1. The coefficient of the linear term is positive.
   The pattern is $(x + ?)(x + ?)$.
   List the positive factors of 8.

Factors of 8		Sum of the factors
1	8	9
2	4	6 ←

2. Find the pair of factors whose sum is 6: 4 and 2.

3. Therefore $x^2 + 6x + 8 = (x + 4)(x + 2)$.

   You can check the result by multiplying $(x + 4)$ and $(x + 2)$.

   $(x + 4)(x + 2) = x^2 + 2x + 4x + 8 = x^2 + 6x + 8 \checkmark$

Example 2   Factor $x^2 - 8x + 15$.

Solution
1. The coefficient of the linear term is negative.
   The pattern is $(x - ?)(x - ?)$
   List the pairs of negative factors of 15.

Factors of 15		Sum of the factors
−1	−15	−16
−3	−5	−8 ←

2. Find the pair of factors whose sum is −8: −3 and −5.

3. Therefore $x^2 - 8x + 15 = (x - 3)(x - 5)$.

**Factor. Check by multiplying the factors. If the polynomial is not
factorable, write** *prime***.**

1. $x^2 + 4x + 3$
2. $x^2 + 8x + 7$
3. $c^2 - 9c + 14$
4. $y^2 - 8y + 12$
5. $r^2 - 5r + 6$
6. $p^2 - 13p + 12$
7. $q^2 + 15q + 14$
8. $n^2 + 9n + 14$
9. $a^2 - 13a + 22$
10. $s^2 - 12s + 30$
11. $x^2 + 18x + 32$
12. $x^2 - 15x + 26$

AME _____   DATE _____

Example 3   Factor $y^2 - 10y + 16$.

Solution   1. Since -10 is negative, think of the negative factors of 16 in your head. (After a little practice you will not need to write all the factors down.)

2. Select the factors of 16 with sum -10: -2 and -8.

3. Therefore $y^2 - 10y + 16 = (y - 2)(y - 8)$.

actor. Check by multiplying the factors. If the polynomial is not
actorable, write *prime*.

3. $a^2 + 10a + 30$

5. $k^2 - 21k + 54$

7. $k^2 - 10k + 21$

9. $k^2 + 7k + 12$

1. $a^2 - 11a + 20$

3. $72 - 17z + z^2$

5. $54 - 15a + a^2$

14. $x^2 - 19x + 60$

16. $n^2 + 23n + 90$

18. $x^2 - 14x + 45$

20. $x^2 - 16x + 48$

22. $x^2 + 22x + 72$

24. $20 - 12c + c^2$

26. $63 - 16c + c^2$

Example 4   Factor $x^2 - 12xy + 32y^2$.

Solution $x^2 - 12xy + 32y^2 = (x - ?)(x - ?)$ Write the factoring pattern.
$= (x - 4y)(x - 8y)$ Fill in the negative factors of $32y^2$.

actor. Check by multiplying the factors. If the polynomial is not
actorable, write *prime*.

7. $x^2 - 11xy + 28y^2$

9. $c^2 - 18cd + 45d^2$

1. $c^2 - 14cd + 24d^2$

3. $y^2 - 16yz + 48z^2$

5. $d^2 + 10de + 24e^2$

28. $a^2 - 9ab + 18b^2$

30. $x^2 - 10xy + 21y^2$

32. $x^2 + 11xy + 30y^2$

34. $a^2 - 18ab + 45b^2$

36. $y^2 - 27yz + 72z^2$

Mixed Review Exercises

olve.

1. $-12 + x = -7$

4. $a + 3 = |2 - 9|$

7. $-\frac{1}{3}x = 9$

2. $d + (-4) = -9$

5. $17m = 68$

8. $\frac{r}{2} - 3 = 6$

3. $-12 + b = 13$

6. $3p + 15 = -60$

9. $-18x = 162$

5–8 Factoring Pattern for x² + bx + c, c negative

Objective: To factor quadratic trinomials whose quadratic coefficient is 1 and whose constant term is negative.

Patterns

Factoring pattern for $x^2 + bx + c$ when c is negative: $(x + ?)(x - ?)$

Example 1   Factor $x^2 - x - 12$.

Solution

1. List the factors of -12 by writing them down or reviewing them mentally.

2. Find the pair of factors with sum -1: 3 and -4.

3. Therefore $x^2 - x - 12 = (x + 3)(x - 4)$.

Factors of −12		Sum of the factors
1	−12	−11
−1	12	11
2	−6	−4
−2	6	4
3	−4	−1 ←
−3	4	1

Example 2   Factor $a^2 + 12a - 45$.

Solution

1. The factoring pattern is $(a + ?)(a - ?)$.

2. Find the pair of factors of -45 whose sum is 12: 15 and -3.

3. Therefore $a^2 + 12a - 45 = (a + 15)(a - 3)$.

Factors of −45		Sum of the factors
1	−45	−44
−1	45	44
3	−15	−12
−3	15	12 ←
5	−9	−4
−5	9	4

Factor. Check by multiplying the factors. If the polynomial is not factorable, write *prime*.

1. $a^2 - 2a - 3$

2. $x^2 + x - 6$

3. $y^2 + 3y - 4$

4. $b^2 - 3b - 10$

5. $c^2 - 9c - 8$

6. $r^2 + 12r - 28$

7. $x^2 - 7x - 18$

8. $y^2 + 4y - 21$

9. $a^2 + 5a - 14$

10. $k^2 - 6k - 40$

11. $z^2 + 6z - 27$

12. $r^2 - 2r - 35$

13. $p^2 - 4p - 12$

14. $a^2 - 3a - 40$

15. $y^2 - 8y - 20$

16. $z^2 - z - 56$

17. $y^2 - 14y - 72$

18. $t^2 + 16t - 30$

–8 Factoring Pattern for $x^2 + bx + c$, c negative (continued)

Example 3 Factor $x^2 - 5kx - 24k^2$.

Solution
1. The factoring pattern is $(x + \text{?})(x - \text{?})$.
2. Find the pair of factors of $-24k^2$ with a sum of $-5k$: $3k$ and $-8k$.
3. Therefore $x^2 - 5kx - 24k^2 = (x + 3k)(x - 8k)$.

Factor. Check by multiplying the factors. If the polynomial is not factorable, write *prime*.

19. $a^2 - ab - 20b^2$ **20.** $y^2 - yz - 12z^2$ **21.** $u^2 - 3uv - 18v^2$

22. $a^2 - 5ab - 24b^2$ **23.** $x^2 - 7xy - 30y^2$ **24.** $h^2 - 2hk - 24k^2$

25. $x^2 + 5xy - 50y^2$ **26.** $c^2 - 2cd - 35d^2$ **27.** $x^2 - 11xy - 42y^2$

Example 4 Factor $1 - 8x - 20x^2$.

Solution Find the pair of factors of $-20x^2$ whose sum is $-8x$: $2x$ and $-10x$.

$$1 - 8x - 20x^2 = (1 + 2x)(1 - 10x)$$

Factor. Check by multiplying the factors. If the polynomial is not factorable, write *prime*.

28. $1 + 2c - 24c^2$ **29.** $1 + 9c - 36c^2$ **30.** $1 + 5x - 24x^2$

31. $1 + 5x - 36x^2$ **32.** $1 - 14y - 72y^2$ **33.** $1 - 12x - 45x^2$

34. $1 - 4x - 21x^2$ **35.** $1 - 7x - 30x^2$ **36.** $1 + 7x - 44x^2$

Mixed Review Exercises

Simplify.

1. $(8x^2y)(4xy^2)(3y^2)$ **2.** $(3x - 2)(2x + 3)$ **3.** $-5x(2x^2 - x + 3)$

4. $(2x - 3)^2$ **5.** $(5x^4y^2)^3$ **6.** $4y(2y^2 + 5y + 3)$

7. $\dfrac{4(xy)^4}{8(xy)^2}$ **8.** $\dfrac{-3ab}{-18a^2b^3}$ **9.** $\dfrac{(-n)^4}{-n^8}$

10. $(m + 2n)^2$ **11.** $(a - 4)(3a + 2)$ **12.** $(2y + 5)^2$

Factor.

13. $10m - 14n + 2$ **14.** $81k^2 - 25$ **15.** $a^2 + 8a + 16$

16. $a^2 - 11ab + 24b^2$ **17.** $18x^2 + 12x$ **18.** $49 - n^2$

19. $u^2 - 18u + 81$ **20.** $27 + 12y + y^2$ **21.** $6a^3b^2 - 18a^2b$

22. $25w^4 - 9x^2$ **23.** $m^2 + 3m + 2$ **24.** $c^2 - 9c - 22$

5–9 Factoring Pattern for ax² + bx + c

Objective: To factor general quadratic trinomials with integral coefficients.

Patterns

Factoring pattern for $ax^2 + bx + c$: $(px + r)(qx + s)$.

Example 1   Factor $2x^2 - 3x - 9$.

Solution

Clue 1   Because the trinomial has a negative constant term, one of r and s will be negative and the other will be positive.

Clue 2   You can list the possible factors of the quadratic term, $2x^2$, and the possible factors of the constant term, -9.

Factors of $2x^2$	Factors of -9	
$2x, x$	1, −9	−1, 9
	3, −3	−3, 3
	9, −1	−9, 1

Make a chart to test the possibilities to see which produces the correct linear term, $-3x$.

Since $(2x + 3)(x - 3)$ gives the correct linear term,
$2x^2 - 3x - 9 = (2x + 3)(x - 3)$.

Possible factors	Linear Term
$(2x + 1)(x - 9)$	$(-18 + 1)x = -17x$
$(2x + 3)(x - 3)$	$(-6 + 3)x = -3x$ ←
$(2x + 9)(x - 1)$	$(-2 + 9)x = 7x$
$(2x - 1)(x + 9)$	$(18 - 1)x = 17x$
$(2x - 3)(x + 3)$	$(6 - 3)x = 3x$
$(2x - 9)(x + 1)$	$(2 - 9)x = -7x$

Example 2   Factor $10x^2 - 11x + 3$.

Solution

Clue 1   Because the trinomial has a positive constant term and a negative linear term, both r and s will be negative.

Clue 2   List the factors of the quadratic term, $10x^2$, and the negative factors of the constant term, 3.

Factors of $10x^2$	Factors of 3
$x, 10x$	−3, −1
$2x, 5x$	−1, −3

Test the possibilities to see which produces $-11x$. Since $(2x - 1)(5x - 3)$ gives the correct linear term, $10x^2 - 11x + 3 = (2x - 1)(5x - 3)$.

Possible factors	Linear term
$(x - 3)(10x - 1)$	$(-1 - 30)x = -31x$
$(x - 1)(10x - 3)$	$(-3 - 10)x = -13x$
$(2x - 3)(5x - 1)$	$(-2 - 15)x = -17x$
$(2x - 1)(5x - 3)$	$(-6 - 5)x = -11x$ ←

Factor. Check by multiplying the factors. If the polynomial is not factorable, write *prime*.

1. $2x^2 + 5x + 2$
2. $2n^2 - 7n + 3$
3. $5y^2 - 9y - 2$
4. $3a^2 + 7a + 2$

5. $4y^2 - 5y + 1$
6. $2a^2 + 11a + 5$
7. $5a^2 - 11a + 2$
8. $7y^2 - 9y + 2$

–9 Factoring Pattern for $ax^2 + bx + c$ (continued)

Factor. Check by multiplying the factors. If the polynomial is not factorable, write *prime*.

9. $2k^2 - 5k - 1$ **10.** $12k^2 - 8k + 1$ **11.** $4x^2 + 17x - 15$ **12.** $2a^2 + 7a + 5$

13. $8y^2 + 6y - 9$ **14.** $9x^2 + 3x - 2$ **15.** $7k^2 - 11k - 6$ **16.** $4u^2 - 8u - 5$

Example 3   Factor $5 - 7x - 6x^2$.

Solution $\begin{aligned} 5 - 7x - 6x^2 &= -6x^2 - 7x + 5 \\ &= (-1)(6x^2 + 7x - 5) \\ &= (-1)(2x - 1)(3x + 5) \\ &= -(2x - 1)(3x + 5) \end{aligned}$ Arrange the terms by decreasing degree.
Factor -1 from each term.
Factor the resulting trinomial.

Note: If you factor $5 - 7x - 6x^2$ directly, you will get $(5 + 3x)(1 - 2x)$.
Since $(1 - 2x) = -(2x - 1)$, the two answers are equivalent.

Factor. Check by multiplying the factors. If the polynomial is not factorable, write *prime*.

17. $10 - 9y - 2y^2$ **18.** $10 - x - 3x^2$ **19.** $3 - x - 10x^2$

20. $3 - 7x - 6x^2$ **21.** $10 - u - 2u^2$ **22.** $5 + 8x - 4x^2$

Example 4   Factor $5a^2 + 2ab - 7b^2$.

Solution $\begin{aligned} 5a^2 + 2ab - 7b^2 &= (a\quad)(5a\quad) \\ &= (a - ?)(5a + ?) \\ &= (a - b)(5a + 7b) \end{aligned}$ Write the factors of $5a^2$.
Test possibilities.

Note: If you write $(a + ?)(5a - ?)$ as the second step, you will not find a combination of factors that produces the desired linear term.

Factor. Check by multiplying the factors.

23. $x^2 - xy - 20y^2$ **24.** $4a^2 - 4ab - 3b^2$ **25.** $3a^2 - 5ab - 12b^2$

26. $5a^2 + 2ab - 7b^2$ **27.** $2x^2 - xy - 3y^2$ **28.** $8y^2 - 6yz - 9z^2$

Mixed Review Exercises

Factor.

1. $x^2 - 196$ **2.** $x^2 - 7x + 12$ **3.** $r^2 - 5r - 36$

4. $c^2 - 10c + 25$ **5.** $9y^2 - 121x^2$ **6.** $4a^2 - 25$

7. $y^2 + 13y + 36$ **8.** $p^2 + 14p + 49$ **9.** $9y^2 + 12y + 4$

10. $m^2 - m - 56$ **11.** $n^2 + 13n + 36$ **12.** $b^2 - 3b - 54$

5–10 Factoring by Grouping

Objective: To factor a polynomial by grouping terms.

Example 1   Factor:
  a. $3(x - y) + w(x - y)$
  b. $m(m + 3n) - (m + 3n)$
  c. $r(p - q) + s(p - q) + t(p - q)$

Solution   Use the distributive property: $ba + ca = (b + c)a$.

This property is valid when a represents any polynomial. For example:

If $a = x - y$, you have $b(x - y) + c(x - y) = (b + c)(x - y)$.

  a. $3(x - y) + w(x - y) = (3 + w)(x - y)$

  b. $m(m + 3n) - (m + 3n) = m(m + 3n) - 1(m + 3n)$
         $= (m - 1)(m + 3n)$

  c. $r(p - q) + s(p - q) + t(p - q) = (r + s + t)(p - q)$

Factor.

1. $2(x + y) + z(x + y)$

2. $5(a - b) + c(a - b)$

3. $e(f + g) - 4(f + g)$

4. $w(x - y) - 6(x - y)$

5. $(c + 2d) - e(c + 2d)$

6. $2c(a - b) - (a - b)$

7. $2x(m - n) - (m - n)$

8. $r(p - q) - (p - q)$

9. $3u(u - 2v) + v(u - 2v) + (u - 2v)$

10. $c(a + b) - d(a + b) + e(a + b)$

Example 2   Factor $7(a - 2) - a(2 - a)$.

Solution   Notice that $a - 2$ and $2 - a$ are opposites.

$7(a - 2) - a(2 - a) = 7(a - 2) - a[-(a - 2)]$ Write $-(a - 2)$ for $2 - a$.
$= 7(a - 2) + a(a - 2)$ Use the distributive property.
$= (7 + a)(a - 2)$

Check: $(7 + a)(a - 2) = 7a - 14 + a^2 - 2a$
$= (7a - 14) + (a^2 - 2a)$
$= 7(a - 2) - (2a - a^2)$
$= 7(a - 2) - a(2 - a)$ $\checkmark$

Therefore, $7(a - 2) - a(2 - a) = (7 + a)(a - 2)$.

Factor. Check by multiplying the factors.

11. $2x(m - n) - (n - m)$

12. $w(x - y) - 7(y - x)$

13. $6(r - s) + t(s - r)$

14. $6(m - n) + p(n - m)$

15. $u(v - 3) + 3(3 - v)$

16. $3x(x - y) + y(y - x)$

17. $x(x - 5) - (5 - x)$

18. $h(h - 6) - 2(6 - h)$

5–10 Factoring by Grouping (continued)

Example 3   Factor $ax - 2x + ay - 2y$.

Solution 1
$$
\begin{aligned}
ax - 2x + ay - 2y &= (ax - 2x) + (ay - 2y) && \text{Group terms with common factors.} \\
&= x(a - 2) + y(a - 2) && \text{Factor each group of terms.} \\
&= (x + y)(a - 2) && \text{Use the distributive property.}
\end{aligned}
$$

Solution 2
$$
\begin{aligned}
ax - 2x + ay - 2y &= (ax + ay) - (2x + 2y) && \text{Group terms with common factors.} \\
&= a(x + y) - 2(x + y) && \text{Factor each group of terms.} \\
&= (a - 2)(x + y) && \text{Use the distributive property.}
\end{aligned}
$$

Factor. Check by multiplying the factors.

19. $2a + ab + 2c + bc$

20. $rs - 6r + st - 6t$

21. $x^2 - 3x + xy - 3y$

22. $u^2 + 3u + uv + 3v$

23. $xy - xz - 3y + 3z$

24. $5t - 10 - st + 2s$

25. $mx + m + 3x + 3$

26. $5x - 5y + wx - wy$

27. $5m^3 - 3m^2 + 10m - 6$

28. $2a^3 + a^2 - 6a - 3$

29. $a^2 - 3ab + ac - 3bc$

30. $2ab - b - 4a + 2$

31. $2u^3 - u^2 - 4u + 2$

32. $x^3 - 4x^2 - x + 4$

Example 4   Factor $(a + 2b)^2 - c^2$ as a difference of two squares.

Solution
$$
\begin{aligned}
(a + 2b)^2 - c^2 &= [(a + 2b) + c][(a + 2b) - c] \\
&= (a + 2b + c)(a + 2b - c)
\end{aligned}
$$
$\left\{ \begin{array}{l} \text{Use the pattern} \\ a^2 - b^2 = (a + b)(a - b). \end{array} \right.$

Factor as a difference of squares.

33. $(a - b)^2 - 4c^2$

34. $(x + 3y)^2 - 16z^2$

35. $x^2 - (y + z)^2$

36. $9p^2 - (q - 2r)^2$

37. $m^2 - (n + 3)^2$

38. $h^2 - (k - 6)^2$

39. $m^2 - (n - 1)^2$

40. $4(x - y)^2 - 25$

Mixed Review Exercises

Solve.

1. $-10 + x = -27$

2. $-n + 8 = 3$

3. $16 + x = 34$

4. $13 = 1 + 3x$

5. $9m - 6m = 27$

6. $4n - 2n + 6 = 12$

7. $12x = 600$

8. $-11m = 143$

9. $7b = 105$

10. $9n = 3n - 30$

11. $17m = 44 + 13m$

12. $9y + 3 = 3(17 - y)$

5–11 *Using Several Methods of Factoring*

Objective: To factor polynomials completely.

Vocabulary

Factored completely A polynomial expressed as the product of a monomial and one or more prime polynomials, that is when it cannot be factored further.

Guidelines for Factoring Completely

1. Factor out the greatest monomial factor first.

2. Look for a difference of squares.
   Pattern: $a^2 - b^2 = (a - b)(a + b)$ (However, $a^2 + b^2$ can't be factored.)

3. Look for a perfect square trinomial.
   Patterns: $(a + b)^2 = a^2 + 2ab + b^2$ $(a - b)^2 = a^2 - 2ab + b^2$

4. If the trinomial is not a perfect square, look for a pair of binomial factors.

5. If a polynomial has four or more terms, look for a way to group the terms in pairs or in a group of three terms that is a perfect square trinomial.

6. Make sure that each binomial or trinomial factor is prime.

7. Check your work by multiplying the factors.

Example 1   Factor $8x^3 - 512x$ completely.

Solution $8x^3 - 512x = 8x(x^2 - 64)$

Greatest monomial factor ⌐↑   ↑⌐ Difference of squares

                   $= 8x(x + 8)(x - 8)$

Example 2   Factor $3x^3 + 3x^2 - 18x$ completely.

Solution $3x^3 + 3x^2 - 18x = 3x(x^2 + x - 6)$

Greatest monomial factor ⌐↑   ↑⌐ Trinomial

                   $= 3x(x + 3)(x - 2)$

Factor completely.

1. $3x^3 - 12x$ 2. $5m^3 - 45m$

3. $3a^2 + 6ab + 3b^2$ 4. $-x^3 + 4xy^2$

5. $-12z^3 + 30z^2 + 18z$ 6. $16r^4 - 24r^3 + 9r^2$

7. $20x^3 - 28x^2 + 8x$ 8. $t^3 + t^2 - 2t$

9. $2x^2 - 128$ 10. $2x^4 - 162$

11. $25z^3 - 36y^2z$ 12. $6x^2 + 22xy - 8y^2$

-11 Using Several Methods of Factoring (continued)

Example 3 Factor $5a^2b^3 + 2a^3b^2 - 3ab^4$ completely.

Solution First rewrite the polynomial in order of decreasing degree in a.

$$5a^2b^3 + 2a^3b^2 - 3ab^4 = 2a^3b^2 + 5a^2b^3 - 3ab^4$$
$$= ab^2(\underbrace{2a^2 + 5ab - 3b^2})$$

Greatest monomial factor ⌐────────────── Trinomial

$$= ab^2(2a - b)(a + 3b)$$

Example 4 Factor $a^2b - 4b + 3a^2 - 12$ completely.

Solution $a^2b - 4b + 3a^2 - 12 = b(a^2 - 4) + 3(a^2 - 4)$ Group and factor.
$$= (b + 3)(\underbrace{a^2 - 4})$$ Use the distributive property.

────────────Difference of squares

$$= (b + 3)(a + 2)(a - 2)$$

Factor completely.

. $a^3x - 9ax^3$

. $20 - 60x + 45x^2$

. $9x^3 + 108x + 63x^2$

. $32r^4 - 48r^3 + 18r^2$

. $bc^2 - 4b + 3c^2 - 12$

. $x^2 + 6xy + 9y^2 - 16$

. $y^4 - 9y^2 + 20$

. $x^4 - 13x^2 + 36$

. $b^4 - 8b^2 + 16$

14. $18x^3 - 24x^2 + 8x$

16. $6x^2 - 18xy + 12y^2$

18. $10k^3 + 25k - 35k^2$

20. $12ab - 3b^2 - 12a^2$

22. $x^3 - x + 6x^2 - 6$

24. $a^3 + a^2b - ab^2 - b^3$

26. $x^4 - 10x^2 + 9$

28. $x^4 - 24x^2 + 144$

30. $a^3 + 2a^2 - 5a - 10$

Mixed Review Exercises

Simplify.

. $\left(-\dfrac{1}{3}\right)\left(\dfrac{1}{4}\right)(60)$

2. $\dfrac{1}{8}(56)$

3. $-\dfrac{1}{7}(56)\left(-\dfrac{1}{8}\right)$

. $\dfrac{120b}{8}$

5. $45 \div \left(\dfrac{1}{5}\right)$

6. $600 \div (-5)$

Factor.

. $x^2 - 11x + 30$

8. $x^2 + 2x - 35$

9. $x^2 - x - 20$

. $2n^2 + 15n + 7$

11. $3x^2 + 7x + 4$

12. $(2x - 6) - 3n(3 - x)$

5–12 *Solving Equations by Factoring*

Objective: To use factoring in solving polynomial equations.

Vocabulary

Zero-product property A product of factors is zero if and only if one or more of the factors is zero.

Polynomial equation An equation whose sides are both polynomials.

Linear equation A polynomial equation whose term of highest degree has degree 1. For example, $x - 2 = 0$ and $5x - 4 = 6$.

Quadratic equation A polynomial equation whose term of highest degree has degree 2. For example, $x^2 - x - 6 = 0$, $x^2 = 9x$, and $10x - 9 = x^2$.

Cubic equation A polynomial equation whose term of highest degree has degree 3. For example, $x^3 - 2x^2 + x - 1 = 0$.

Standard form of a polynomial equation A form of an equation in which one side is a simplified polynomial arranged in order of decreasing degree of the variable and the other side is zero.

Double or multiple root A factor that occurs twice in the factored form of an equation. For example, 5 is a double root of $x(x - 5)(x - 5) = 0$.

Example 1 Solve $(x - 1)(x + 3) = 0$.

Solution Since the product of factors is 0, one of the factors on the left side must equal 0.

$$x - 1 = 0 \quad \text{or} \quad x + 3 = 0$$
$$x = 1 \qquad\qquad x = -3$$

The solution set is $\{1, -3\}$. Just by looking at the original equation, you can see that when $x = 1$ or $x = -3$, the product will be 0.

Example 2 Solve $3n(n - 2)(n - 5) = 0$.

Solution $3n = 0 \quad$ or $\quad n - 2 = 0 \quad$ or $\quad n - 5 = 0$
$n = 0 \qquad\qquad n = 2 \qquad\qquad n = 5 \quad$ The solution set is $\{0, 2, 5\}$.

CAUTION Never transform an equation by dividing by an expression containing a variable. Notice that in Example 2, the solution 0 would have been lost if both sides of the equation had been divided by $3n$.

Solve.

1. $(y + 4)(y - 5) = 0$ **2.** $0 = (n + 1)(n + 8)$ **3.** $10n(n - 2) = 0$

4. $2x(x - 10) = 0$ **5.** $(p - 1)(p - 7) = 0$ **6.** $0 = 2n(n - 1)(n - 3)$

7. $x(2x - 1)(2x + 1) = 0$ **8.** $0 = n(n - 6)$ **9.** $0 = 3x(4x - 1)(x - 2)$

–12 Solving Equations by Factoring (continued)

Example 3   Solve the quadratic equation $2x^2 - x = 3$.

Solution
1. Transform the equation into standard form. $2x^2 - x - 3 = 0$
2. Factor the left side. $(2x - 3)(x + 1) = 0$
3. Set each factor equal to 0 and solve.

$$2x - 3 = 0 \quad \text{or} \quad x + 1 = 0$$
$$2x = 3 \qquad\qquad x = -1$$
$$x = \frac{3}{2}$$

4. Check the solutions in the original equation.

$$2\left(\frac{3}{2}\right)^2 - \left(\frac{3}{2}\right) \overset{?}{=} 3 \qquad\qquad 2(-1)^2 - (-1) \overset{?}{=} 3$$
$$2\left(\frac{9}{4}\right) - \frac{3}{2} \overset{?}{=} 3 \qquad\qquad 2(1) + 1 \overset{?}{=} 3$$
$$\frac{9}{2} - \frac{3}{2} \overset{?}{=} 3 \qquad\qquad 3 = 3 \checkmark$$
$$\frac{9}{2} - \frac{3}{2} = \frac{6}{2} = 3 \checkmark$$

The solution set is $\left\{-1, \dfrac{3}{2}\right\}$.

Solve.

10. $x^2 - x - 12 = 0$
11. $x^2 - 12x + 27 = 0$
12. $0 = x^2 - 4x - 32$

13. $0 = m^2 + 3m - 54$
14. $x^2 - 4y + 3 = 0$
15. $x^2 - 10x - 24 = 0$

16. $0 = n^2 - n$
17. $y^2 = 12y$
18. $6k^2 = 2k$

19. $x^2 + 16 = 8x$
20. $a^2 = 10 - 3a$
21. $3x^2 - x = 2$

22. $0 = x^2 + 12x + 35$
23. $y^2 + 5y = 14$
24. $x^2 = 5x + 36$

25. $4m^2 - 25 = 0$
26. $r^2 + 8 = 9r$
27. $6n^2 - n = 2$

28. $3x^2 + 1 = 4x$
29. $3a^2 = 6a$
30. $3p^2 - 14p = 80$

31. $2x^2 = 10 + x$
32. $3p^2 + 17p = -10$
33. $3x^2 + 1 = 4x$

Mixed Review Exercises

Evaluate if $x = 3$ and $y = 6$.

1. $(x - y)^3$
2. $x^3 \cdot x^2$
3. $4x^3$

4. $(4x)^3$
5. $3x + y^2$
6. $3x^2 + y$

7. $3(x + y)^2$
8. $(yx)^2$
9. y^2x^2

Simplify.

10. $(5x^2y^2)(-3xy^4)$
11. $(8a)^3$
12. $-3(x + 4)$

5–13 Using Factoring to Solve Problems

Objective: To solve problems by writing and factoring quadratic equations.

Example 1   Find two consecutive positive odd integers whose product is 143.

Solution

Step 1   The problem asks for two consecutive positive odd integers.

Step 2   Let n = the first integer. Then $n + 2$ = the second integer.

Step 3   Use the facts in the problem to write an equation. $n(n + 2) = 143$

Step 4   Solve the equation.

$$n^2 + 2n - 143 = 0$$
$$(n + 13)(n - 11) = 0$$

$$n + 13 = 0 \quad \text{or} \quad n - 11 = 0$$
$$n = -13 \qquad\qquad n = 11$$

You are to find positive odd integers, so reject -13. If $n = 11$, then $n + 2 = 13$.

Step 5   *Check:* $11 \times 13 = 143$.   The integers are 11 and 13.

Example 2   Originally a rectangle was 8 cm by 17 cm. When both dimensions were decreased by the same amount, the area of the rectangle decreased by 66 cm^2. Find the dimensions of the new rectangle.

Solution

Step 1   The problem asks for the dimensions of the new rectangle.

Step 2   Let x = the amount by which each dimension is decreased. Make a sketch. The new dimensions are $17 - x$ and $8 - x$.

Step 3 $\left(\begin{array}{c}\text{Original}\\ \text{area}\end{array}\right) - \left(\begin{array}{c}\text{Decrease}\\ \text{in area}\end{array}\right) = \left(\begin{array}{c}\text{New}\\ \text{area}\end{array}\right)$

$\quad (17 \cdot 8) \quad - \quad\quad 66 \quad\quad = \quad (17 - x)(8 - x)$

Step 4
$$136 - 66 = 136 - 25x + x^2$$
$$70 = 136 - 25x + x^2$$
$$0 = 66 - 25x + x^2$$
$$0 = x^2 - 25x + 66$$
$$0 = (x - 3)(x - 22)$$

$$x - 3 = 0 \quad \text{or} \quad x - 22 = 0$$
$$x = 3 \qquad\qquad x = 22$$

Step 5   Check in the words of the problem and you'll see that you must reject 22.

The new rectangle is 14 cm long and 5 cm wide.

CAUTION   A solution of an equation may not satisfy some of the conditions of the problem. You reject solutions of an equation that do not make sense for the problem.

–13 *Using Factoring to Solve Problems* (continued)

Solve.

1. If a number is added to its square, the result is 72. Find the number.

2. If a number is subtracted from its square, the result is 90. Find the number.

3. A positive number is 56 less than its square. Find the number.

4. A negative number is 56 less than its square. Find the number.

5. Find two consecutive negative integers whose product is 72.

6. Find two consecutive positive even integers whose product is 120.

7. The sum of the squares of two consecutive positive odd integers is 202. Find the integers.

8. The sum of the squares of two consecutive negative odd integers is 130. Find the integers.

9. The length of a rectangle is 5 cm greater than its width. Find the dimensions of the rectangle if its area is 126 cm^2.

10. The length of a rectangle is 8 cm less than twice its width. Find the dimensions of the rectangle if the area is 120 cm^2.

11. Find the dimensions of a rectangle whose perimeter is 40 m and whose area is 96 m^2. (*Hint:* Let the width be w. Use the perimeter to find the length in terms of w.)

12. Find the dimensions of a rectangle whose perimeter is 52 m and whose area is 160 m^2.

13. Originally the dimensions of a rectangle were 12 cm by 7 cm. When both dimensions were decreased by the same amount, the area of the rectangle decreased by 34 cm^2. Find the dimensions of the new rectangle.

14. Originally a rectangle was twice as long as it was wide. When 5 m was subtracted from its length and 3 m subtracted from its width, the resulting rectangle had an area of 55 m^2. Find the dimensions of the new rectangle.

Mixed Review Exercises

Simplify.

1. $(6ab^2)(3a^2b)$

2. $(4a^2)^3$

3. $2a(3 - 2a)$

4. $(8r)\left(\frac{1}{4}rs^2\right)$

5. $(2by^2)^2$

6. $\left(\frac{1}{6}\right)(18n - 30p)$

7. $(2a + 3)(2a^2 + 3 - 5a)$

8. $(-20x - 15y)\left(-\frac{1}{5}\right)$

9. $(5m - 3)^2$

Factor completely.

10. $6m^2 + 2m - 20$

11. $28a^3 - 7ab^2$

12. $15n^2 + 14n - 8$

13. $y^4 - y^3 - 6y^2$

14. $12m^2 + 13mn + 3n^2$

15. $8x^2 - 14x + 3$

6 Fractions

6–1 *Simplifying Fractions*

Objective: To simplify algebraic fractions.

Vocabulary

Simplest form of an algebraic fraction A form of the fraction in which the numerator and denominator have no common factor other than 1 and −1.

CAUTION In a fraction, you cannot cancel terms. You must factor to find common factors to cancel. For example, $\dfrac{x + y}{x + 2} \neq \dfrac{y}{2}$.

Example 1 Simplify: **a.** $\dfrac{21x - 14y}{7}$ **b.** $\dfrac{3c - 24}{c - 8}$ **c.** $\dfrac{2a + 6}{4a - 12}$

Solution Factor. Then look for common factors to cancel.

a. $\dfrac{21x - 14y}{7} = \dfrac{\cancel{7}(3x - 2y)}{\cancel{7}} = 3x - 2y$

b. $\dfrac{3c - 24}{c - 8} = \dfrac{3\cancel{(c - 8)}}{\cancel{c - 8}} = 3 \ (c \neq 8)$ $\begin{cases} \text{The denominator can't equal 0.} \\ \text{So } c - 8 \neq 0, \text{ or } c \neq 8. \end{cases}$

c. $\dfrac{2a + 6}{4a - 12} = \dfrac{\overset{1}{\cancel{2}}(a + 3)}{\underset{2}{\cancel{4}}(a - 3)}$

$= \dfrac{1(a + 3)}{2(a - 3)}$

$= \dfrac{a + 3}{2(a - 3)}, \quad (a \neq 3)$ $\begin{cases} \text{If } a = 3, a - 3 = 0. \\ \text{You must restrict the variable} \\ \text{in the denominator.} \end{cases}$

Simplify. Give any restrictions on the variables.

1. $\dfrac{3x - 3y}{9}$ 2. $\dfrac{10m - 15n}{5}$ 3. $\dfrac{4a - 20}{a - 5}$ 4. $\dfrac{3n + 12}{n + 4}$ 5. $\dfrac{4n + 24}{n + 6}$

6. $\dfrac{2n - 18}{n - 9}$ 7. $\dfrac{2m + 3}{6m + 9}$ 8. $\dfrac{6x + 6y}{6x - 6y}$ 9. $\dfrac{3w + 5}{9w + 15}$ 10. $\dfrac{4m - 4n}{4m + 4n}$

Example 2 Simplify $\dfrac{x^2 - 4}{2x^2 + 3x - 2}$.

Solution $\dfrac{x^2 - 4}{2x^2 + 3x - 2} = \dfrac{(x - 2)\cancel{(x + 2)}}{(2x - 1)\cancel{(x + 2)}}$ Factor. $x + 2$ is a common factor.

$= \dfrac{x - 2}{2x - 1}, \left(x \neq -2, x \neq \dfrac{1}{2} \right)$

To see which values of x to exclude, look at the denominator of the original fraction. Since $2x - 1 \neq 0$ and $x + 2 \neq 0$, $x \neq \dfrac{1}{2}$ and $x \neq -2$.

-1 Simplifying Fractions (continued)

Simplify. Give any restrictions on the variables.

11. $\dfrac{3x - 9}{x^2 - 9}$

12. $\dfrac{5y + 30}{y^2 - 36}$

13. $\dfrac{b^2 - 4}{b + 2}$

14. $\dfrac{x^2 - 49}{x + 7}$

15. $\dfrac{8n^2 - 72}{4n - 12}$

16. $\dfrac{15c + 25d}{90c^2 - 250d^2}$

17. $\dfrac{4xy}{x^2y - xy^2}$

18. $\dfrac{3x^2 - 6x}{3x^3}$

19. $\dfrac{a^2 - 3a - 10}{a^2 - 4}$

20. $\dfrac{a^2 - 5a - 36}{a^2 - 81}$

21. $\dfrac{2w^2 - w - 6}{2w - 4}$

22. $\dfrac{2x^2 + 5x - 3}{x^2 + 2x - 3}$

Example 3   Simplify: $\dfrac{2x^2 - 3x - 2}{4 - x^2}$.

Solution $\dfrac{2x^2 - 3x - 2}{4 - x^2} = \dfrac{(2x + 1)(x - 2)}{(2 + x)(2 - x)}$ $\Big\{$ Factor. Since $(x - 2)$ and $(2 - x)$ are opposites, $(2 - x) = -(x - 2)$.

$\qquad\qquad = \dfrac{(2x + 1)\cancel{(x - 2)}}{-(2 + x)\cancel{(x - 2)}}$

$\qquad\qquad = \dfrac{2x + 1}{-(2 + x)}$, or $-\dfrac{2x + 1}{x + 2}$, $(x \neq 2, x \neq -2)$

Simplify. Give any restrictions on the variables.

23. $\dfrac{(3n + 2)(n - 3)}{(3 + n)(3 - n)}$

24. $\dfrac{(x - 4)(3x + 4)}{(4 - x)(5x + 2)}$

25. $\dfrac{(x - 5)(2x - 7)}{(5 - x)(3x + 2)}$

26. $\dfrac{(x - 7)(x - 4)}{(7 - x)(x + 2)}$

27. $\dfrac{x^2 - 10x + 25}{25 - x^2}$

28. $\dfrac{6 - x}{x^2 - 2x - 24}$

29. $\dfrac{(a - 3)^2}{9 - a^2}$

30. $\dfrac{2n^2 - 72}{6n + 36}$

31. $\dfrac{6 + x - x^2}{x^2 - 9}$

32. $\dfrac{10 + 3x - x^2}{x^2 - 4}$

33. $\dfrac{2w^2 - w - 6}{2w - 4}$

34. $\dfrac{3x^2 - 6x}{6x^2 - 7x - 10}$

35. $\dfrac{2n^2 + 5n - 3}{4n^2 + 8n - 5}$

36. $\dfrac{2y^2 - 7y + 3}{6y - 2y^2}$

37. $\dfrac{3y^2 - 5y + 2}{6y^2 - y - 2}$

38. $\dfrac{3x^2 - 15x}{3x^2 - 16x + 5}$

Mixed Review Exercises

Simplify. Assume that no denominator equals zero.

1. $10\left(\dfrac{1}{2}u + \dfrac{1}{5}v\right)$

2. $(-36m + 24n)\left(-\dfrac{1}{6}\right)$

3. $\dfrac{20a^6b^5}{35a^2b^3}$

4. $\dfrac{(-2y)^4}{(y^2)^4}$

5. $\dfrac{2x^4 + 6x^3 + 10x^2}{2x^2}$

6. $(-10)(-6)(-2)(-5)$

Solve.

7. $3(x + 1) + 1 = 25$

8. $8y - (5y + 4) = 11$

9. $(2n - 3) - (5 - 2n) = 16$

6–2 *Multiplying Fractions*

Objective: To multiply algebraic fractions.

Multiplication Rule for Fractions To multiply fractions, you multiply their numerators and multiply their denominators.

$$\frac{a}{b} \cdot \frac{c}{d} = \frac{ac}{bd} \qquad \text{For example,} \qquad \frac{3}{4} \cdot \frac{5}{8} = \frac{3 \cdot 5}{4 \cdot 8} = \frac{15}{32}.$$

Example 1   Multiply: $\frac{5}{6} \cdot \frac{9}{10}$

Solution 1 $\frac{5}{6} \cdot \frac{9}{10} = \frac{5 \cdot 9}{6 \cdot 10} = \frac{45}{60} = \frac{3}{4}$ You can multiply first and then simplify.

Solution 2 $\frac{1\cancel{5}}{2\cancel{6}} \cdot \frac{\cancel{9}3}{\cancel{10}2} = \frac{3}{4}$ You can simplify first and then multiply.

Multiply. Express each product in simplest form.

1. $\frac{3}{7} \cdot \frac{35}{9}$

2. $\frac{5}{16} \cdot \frac{4}{15}$

3. $\frac{12}{7} \cdot \frac{14}{9}$

4. $-\frac{5}{2} \cdot \frac{16}{25}$

5. $\frac{2}{5} \cdot \frac{5}{9} \cdot \frac{9}{10}$

6. $\frac{8}{5} \cdot \frac{3}{4} \cdot \frac{15}{16}$

7. $\left(-\frac{3}{2}\right)^2 \cdot \frac{8}{9}$

8. $(-3)^2 \cdot \frac{25}{12}$

Example 2   a. $\frac{9x}{y^2} \cdot \frac{y^3}{24}$ b. $\frac{x^2 + x - 12}{x^2 + 5x} \cdot \frac{x^2 - 25}{x - 3}$

Solution   a. $\frac{9x}{y^2} \cdot \frac{y^3}{24} = \frac{\cancel{3} \cdot 3x}{\cancel{y^2}} \cdot \frac{\cancel{y^2} \cdot y}{\cancel{3} \cdot 8}$ ← Multiply the numerators.
← Multiply the denominators.

$= \frac{3xy}{8} \ (y \neq 0)$

b. $\frac{x^2 + x - 12}{x^2 + 5x} \cdot \frac{x^2 - 25}{x - 3} = \frac{(x + 4)\cancel{(x - 3)}}{x\cancel{(x + 5)}} \cdot \frac{\cancel{(x + 5)}(x - 5)}{\cancel{(x - 3)}}$

$= \frac{(x + 4)(x - 5)}{x} \ (x \neq 0, \ x \neq -5, \ x \neq 3)$

You can leave the answer in factored form.

CAUTION   From now on, assume that no denominator equals zero. You won't need to show the excluded values, but know what they are.

Multiply. Express each product in simplest form.

9. $\frac{8}{x^2} \cdot \frac{x^3}{4}$

10. $\frac{7y}{5} \cdot \frac{10}{21y}$

11. $\frac{a}{c} \cdot \frac{c}{d} \cdot \frac{d}{e}$

12. $\frac{6}{x^2} \cdot \frac{5x}{12}$

13. $\frac{6w}{v} \cdot \frac{v^3}{3w^2}$

14. $\frac{8a}{11b^3} \cdot \frac{33b}{4a^2}$

15. $\frac{2de^2}{5e^2f} \cdot \frac{f^2}{4d}$

16. $\frac{3rs^2}{4t} \cdot \frac{8t^2}{9rs}$

–2 Multiplying Fractions (continued)

Multiply. Express each product in simplest form.

17. $\dfrac{x-3}{x^2} \cdot \dfrac{2x}{x^2-9}$

18. $\dfrac{x^2-4}{8x} \cdot \dfrac{4x^2}{5x+10}$

19. $\dfrac{a^2-b^2}{a^2} \cdot \dfrac{a}{2(b-a)}$

20. $\dfrac{m}{3(n-m)} \cdot \dfrac{m^2-n^2}{m^2}$

Rule of Exponents for a Power of a Quotient

For every positive integer m, $\left(\dfrac{a}{b}\right)^m = \dfrac{a^m}{b^m}$. For example, $\left(\dfrac{2}{3}\right)^3 = \dfrac{2^3}{3^3}$.

Example 3   Simplify: **a.** $\left(\dfrac{x}{2}\right)^4$ **b.** $\left(-\dfrac{x}{3}\right)^2 \cdot \dfrac{9}{5x}$

Solution   **a.** $\left(\dfrac{x}{2}\right)^4 = \dfrac{x^4}{2^4}$ **b.** $\left(-\dfrac{x}{3}\right)^2 \cdot \dfrac{9}{5x} = \dfrac{x^2}{9} \cdot \dfrac{9}{5x}$

$\qquad\qquad\quad = \dfrac{x^4}{16}$ $\qquad\qquad\qquad\qquad\qquad = \dfrac{x \cdot \cancel{x}}{\cancel{9}} \cdot \dfrac{\cancel{9}}{5 \cdot \cancel{x}}$

$\qquad\qquad\qquad\qquad\qquad\qquad\qquad\qquad\qquad = \dfrac{x}{5}$

Multiply. Express each product in simplest form.

21. $\left(\dfrac{a}{4}\right)^2$

22. $\left(\dfrac{c}{3}\right)^3$

23. $\left(\dfrac{m}{5}\right)^3$

24. $\left(\dfrac{y}{4}\right)^3$

25. $\left(\dfrac{2n}{5}\right)^2$

26. $\left(\dfrac{3x}{4}\right)^2$

27. $\left(\dfrac{3w}{7}\right)^2$

28. $\left(\dfrac{7b}{2}\right)^2$

29. $\left(\dfrac{2a}{3b^3}\right)^2$

30. $\left(\dfrac{4m}{5n^2}\right)^2$

31. $\left(-\dfrac{x}{3y}\right)^2$

32. $-\left(\dfrac{4b^3}{5}\right)^2$

33. $\left(\dfrac{x}{y}\right)^2 \cdot \dfrac{y}{x}$

34. $\left(\dfrac{2x}{y}\right)^3 \cdot \dfrac{y^2}{4}$

35. $\left(-\dfrac{x}{2y}\right)^2\left(-\dfrac{4y}{x}\right)$

36. $\left(\dfrac{2z}{y}\right)^3 \cdot \dfrac{3yz}{8}$

37. Find the area of a square if each side has length $\dfrac{3x}{4}$ in.

38. Find the volume of a cube if each edge has length $\dfrac{2n}{3}$ in.

Mixed Review Exercises

Factor completely.

1. $a^2 + 11a + 24$

2. $x^2 - 5x - 14$

3. $81x^4 - 16$

4. $2x^2 + 5x + 2$

5. $25y^2 - 9z^2$

6. $c^2 + 10c + 25$

7. $xy + 5y - 4xz - 20z$

8. $9x^2 - 6x + 1$

9. $3x^2 - 11x - 4$

10. $x^4 + 6x^2 - 5x^3$

11. $n^2 + 4n - 12$

12. $y^2 - 5y - 36$

6–3 Dividing Fractions

Objective: To divide algebraic fractions.

Division Rule for Fractions To divide by a fraction, you multiply by its reciprocal. Remember that the reciprocal of a number n is the number $\frac{1}{n}$ for which $n \cdot \frac{1}{n} = 1$.

$$\frac{a}{b} \div \frac{c}{d} = \frac{a}{b} \cdot \frac{d}{c} \qquad \text{For example,} \quad \frac{3}{5} \div \frac{2}{7} = \frac{3}{5} \cdot \frac{7}{2} = \frac{21}{10}.$$

Example 1 Divide: $\frac{x}{3y} \div \frac{xy}{9}$

Solution $\frac{x}{3y} \div \frac{xy}{9} = \frac{x}{3y} \cdot \frac{9}{xy}$ Multiply by the reciprocal of $\frac{xy}{9}$.

$$= \frac{\cancel{x}}{\cancel{3} \cdot y} \cdot \frac{\cancel{9} \cdot 3}{\cancel{x} \cdot y}$$ Factor and simplify.

$$= \frac{3}{y^2}$$

Divide. Give your answers in simplest form.

1. $\frac{8}{5} \div \frac{16}{25}$
2. $\frac{3}{4} \div \frac{9}{8}$
3. $\frac{a}{10} \div \frac{a}{2}$
4. $\frac{2x}{5} \div \frac{x}{15}$
5. $\frac{x^2}{y} \div \frac{x}{y^2}$

6. $\frac{4n^2}{5} \div \frac{8n}{25}$
7. $\frac{ab}{4} \div \frac{a}{b}$
8. $\frac{c}{3d} \div \frac{c^2}{9d^2}$
9. $\frac{2x^2}{3y} \div \frac{xy}{9}$
10. $\frac{3n}{4m^2} \div \frac{1}{12mn}$

11. $\frac{x^2y}{2} \div xy$
12. $\frac{8a^2}{3b} \div 4a$
13. $1 \div \left(\frac{2x}{3}\right)^2$
14. $9 \div \left(\frac{3}{n}\right)^2$
15. $16 \div \left(\frac{2}{a}\right)^3$

Example 2 Divide: $\frac{15}{x^2 - 16} \div \frac{20}{x - 4}$

Solution $\frac{15}{x^2 - 16} \div \frac{20}{x - 4} = \frac{15}{x^2 - 16} \cdot \frac{x - 4}{20}$ Multiply by the reciprocal.

$$= \frac{\cancel{5} \cdot 3}{(x + 4)\cancel{(x - 4)}} \cdot \frac{\cancel{x - 4}}{\cancel{5} \cdot 4}$$ Factor and simplify.

$$= \frac{3}{4(x + 4)}$$

Divide. Give your answers in simplest form.

16. $\frac{3 + 3b}{6} \div \frac{1 + b}{9}$
17. $\frac{4n - 2}{8n} \div \frac{2n - 1}{24}$
18. $\frac{x^2 - 4}{3} \div \frac{x + 2}{9}$

19. $\frac{x^2 - 16}{3x} \div \frac{x - 4}{6}$
20. $\frac{x^2 - 9}{3} \div \frac{x + 3}{6}$
21. $\frac{x^2 - 25}{4x} \div \frac{x - 5}{12}$

–3 Dividing Fractions (continued)

Divide. Give your answers in simplest form.

2. $\dfrac{2}{x-3} \div \dfrac{2}{3-x}$

23. $\dfrac{4}{6-3a} \div \dfrac{6}{8-4a}$

24. $\dfrac{x^2+2x}{x^2-4} \div \dfrac{x+2}{x-2}$

5. $\dfrac{1}{3a-12} \div \dfrac{1}{2a-8}$

26. $\dfrac{3x-3y}{x} \div \dfrac{x^2-y^2}{x^2}$

27. $\dfrac{4}{n^2-16} \div \dfrac{8n-32}{n+4}$

Example 3   Divide: $\dfrac{x^2-3x-10}{2x-6} \div \dfrac{x^2-4}{x^2+x-6}$

Solution $\dfrac{x^2-3x-10}{2x-6} \div \dfrac{x^2-4}{x^2+x-6} = \dfrac{x^2-3x-10}{2x-6} \cdot \dfrac{x^2+x-6}{x^2-4}$

$$= \frac{(x-5)(x+2)}{2(x-3)} \cdot \frac{(x+3)(x-2)}{(x+2)(x-2)}$$

$$= \frac{(x-5)(x+3)}{2(x-3)} \quad \begin{cases} \text{Stop; no further simpli-} \\ \text{fication is possible.} \end{cases}$$

Divide. Give your answers in simplest form.

. $\dfrac{x^2-9}{x^2-4} \div \dfrac{x^2-x-6}{x^2+x-6}$

29. $\dfrac{x^2+x-20}{5x+25} \div \dfrac{x^2-4x-5}{x^2-25}$

. $\dfrac{x^2-y^2}{x^2+y^2} \div (x-y)$

31. $\dfrac{x^2-x-6}{x^2+2x+1} \div \dfrac{x+2}{x+1}$

. $\dfrac{x^2-3x+2}{x^2+3x+2} \div \dfrac{4x-8}{8x+8}$

33. $\dfrac{x^2-4}{x+2} \div \dfrac{x-2}{x+1}$

. $\dfrac{x^2-25}{x^2-16} \div \dfrac{4x+20}{8x-32}$

35. $\dfrac{4x^2-y^2}{4y^2-x^2} \div \dfrac{2x-y}{2y-x}$

. $\dfrac{x^2-3x+2}{x^2-7x+10} \div \dfrac{x^2-1}{x^2-4x-5}$

37. $\dfrac{x^2-8x+15}{x^2-9x+14} \div \dfrac{x^2-9}{x^2+x-6}$

. $\dfrac{2x^2+7x+3}{2x^2+5x+2} \div \dfrac{x^2-7x-30}{x^2-6x-40}$

39. $\dfrac{x^2+5x-6}{x^2-x-20} \div \dfrac{x^2+2x-3}{x^2-2x-15}$

ixed Review Exercises

lve.

. $3k = 4k - 11$

2. $5p + 10 = 45$

3. $(4b-3) - (3-2b) = 30$

. $\dfrac{1}{3}(9k-6) = 7$

5. $2n^3 - 32n = 0$

6. $2x^2 + x = 3$

ve the prime factorization of each number.

. 225

8. 136

9. 140

10. 1250

6–4 Least Common Denominators

Objective: To express two or more fractions with their least common denominator.

Example 1 Complete: **a.** $\dfrac{2}{3} = \dfrac{?}{15}$ **b.** $\dfrac{5}{2a} = \dfrac{?}{18a^2}$

Solution To write a fraction in a different form, you can multiply the numerator and denominator by the same nonzero number.

a. $\dfrac{2}{3} = \dfrac{?}{15}$ ←————— 3 is multiplied by 5 to get 15.

$\dfrac{2}{3} = \dfrac{2 \cdot 5}{3 \cdot 5} = \dfrac{10}{15}$ ←————— Therefore, multiply 2 by 5 to get 10.

b. $\dfrac{5}{2a} = \dfrac{?}{18a^2}$ ←————— $2a$ is multiplied by $9a$ to get $18a^2$.

$\dfrac{5}{2a} = \dfrac{5 \cdot 9a}{2a \cdot 9a} = \dfrac{45a}{18a^2}$ ←—— Therefore, multiply 5 by $9a$ to get $45a$.

Complete.

1. $\dfrac{2}{3} = \dfrac{?}{18}$ 2. $\dfrac{3}{5} = \dfrac{?}{20}$ 3. $\dfrac{5}{8} = \dfrac{?}{56}$ 4. $\dfrac{2a}{15} = \dfrac{?}{45}$ 5. $\dfrac{x-2}{3} = \dfrac{?}{12}$

6. $\dfrac{2n-3}{5} = \dfrac{?}{15}$ 7. $\dfrac{8}{15x} = \dfrac{?}{90x^2}$ 8. $\dfrac{5}{3a} = \dfrac{?}{9a^3}$ 9. $\dfrac{x}{3y} = \dfrac{?}{18xy}$ 10. $\dfrac{3n}{4m} = \dfrac{?}{12m^2n}$

Example 2 Complete: $\dfrac{2}{x-3} = \dfrac{?}{(x-3)(x+4)}$

Solution $\dfrac{2}{x-3} = \dfrac{?}{(x-3)(x+4)}$ ←——— $(x-3)$ is multiplied by $(x+4)$.

$\dfrac{2}{x-3} = \dfrac{2(x+4)}{(x-3)(x+4)}$ ←——— Therefore, multiply 2 by $(x+4)$.

Complete.

11. $\dfrac{6}{n-1} = \dfrac{?}{(n-1)(n+4)}$ 12. $\dfrac{4}{x+2} = \dfrac{?}{(x+2)(x-2)}$

13. $\dfrac{3}{2x-1} = \dfrac{?}{(2x-1)^2}$ 14. $\dfrac{5y}{x-7} = \dfrac{?}{(x-7)^2}$

15. $\dfrac{7}{x-3} = \dfrac{?}{4x-12}$ 16. $\dfrac{3}{2x+5} = \dfrac{?}{6x+15}$

17. $\dfrac{3}{x+2} = \dfrac{?}{x^2-4}$ 18. $\dfrac{4}{x-1} = \dfrac{?}{x^2-1}$

19. $\dfrac{5}{3-y} = \dfrac{?}{3y-y^2}$ 20. $\dfrac{3x}{2+x} = \dfrac{?}{2x+x^2}$

–4 Least Common Denominator (continued)

Example 3   Find the LCD of $\dfrac{5}{6}$, $\dfrac{7}{20}$, and $\dfrac{8}{42}$.

Solution
1. Factor each denominator into prime numbers.
   $6 = 2 \cdot 3$ $20 = 2^2 \cdot 5$ $42 = 2 \cdot 3 \cdot 7$

2. Greatest power of 2: 2^2
   Greatest power of 3: 3 $2^2 \cdot 3 \cdot 5 \cdot 7 = 420$
   Greatest power of 5: 5
   Greatest power of 7: 7   The LCD is 420.

---

nd the LCD of each group of fractions.

. $\dfrac{1}{4}, \dfrac{5}{6}$ **22.** $\dfrac{1}{2}, \dfrac{3}{8}$ **23.** $\dfrac{3}{2}, \dfrac{2}{5}, \dfrac{1}{4}$ **24.** $\dfrac{2}{3}, \dfrac{5}{9}, \dfrac{1}{6}$ **25.** $\dfrac{5}{8}, \dfrac{2}{5}, \dfrac{4}{3}$ **26.** $\dfrac{2}{3}, \dfrac{3}{4}, \dfrac{5}{9}$

Example 4   Find the LCD of $\dfrac{5}{9x - 36}$ and $\dfrac{4}{5x - 20}$.

Solution
1. Factor each denominator completely. Factor integers into primes.
   $9x - 36 = 9(x - 4) = 3^2(x - 4)$ $5x - 20 = 5(x - 4)$

2. Form the product of the greatest power of each factor.
   $$3^2 \cdot 5(x - 4) = 45(x - 4)$$

   The LCD is $45(x - 4)$.

---

nd the LCD of each group of fractions.

. $\dfrac{a + 2b}{4}, \dfrac{2b - a}{6}$ **28.** $\dfrac{n - 2}{12}, \dfrac{n + 3}{15}$ **29.** $\dfrac{n - 1}{15}, \dfrac{n + 3}{20}$

. $\dfrac{2x + 3}{12}, \dfrac{x - 4}{8}$ **31.** $\dfrac{x + 2y}{25}, \dfrac{2x + y}{20}$ **32.** $\dfrac{x^2 - x - 6}{21}, \dfrac{x^2 - 9}{35}$

. $\dfrac{2}{3t}, \dfrac{5}{9rt^2}$ **34.** $\dfrac{5}{xy}, \dfrac{6}{y^2}$ **35.** $\dfrac{11}{m^2n}, \dfrac{17}{mn^2}$

. $\dfrac{3}{2x + 10}, \dfrac{x}{5x + 25}$ **37.** $\dfrac{3a}{a + 1}, \dfrac{2}{a - 1}$ **38.** $\dfrac{3}{a^2 - 4}, \dfrac{5}{a + 2}$

. $\dfrac{x}{x^2 + 3x}, \dfrac{2x}{x^2 - 3x}$ **40.** $\dfrac{7}{n + 3}, \dfrac{n - 1}{n^2 + n - 6}$ **41.** $\dfrac{a + 1}{a - 2}, \dfrac{a - 5}{a^2 - 5a + 6}$

---

ixed Review Exercises

ctor completely.

. $3n - 9q + 15$ **2.** $2x^2 - 8$ **3.** $x^2 - 10x + 16$

. $x^2 - 5x - 36$ **5.** $2x^2 - 5x - 3$ **6.** $x^2 + 14x + 24$

. $x^2 - 4x - 32$ **8.** $x^2 + 24x + 144$ **9.** $n^2 - 6n$

6-5 Adding and Subtracting Fractions

Objective: To add and subtract algebraic fractions.

Rules for Fractions

Addition Rule for Fractions

$$\frac{a}{c} + \frac{b}{c} = \frac{a + b}{c}$$ For example, $\frac{3}{5} + \frac{1}{5} = \frac{3 + 1}{5} = \frac{4}{5}$

Subtraction Rule for Fractions

$$\frac{a}{c} - \frac{b}{c} = \frac{a - b}{c}$$ For example, $\frac{5}{9} - \frac{1}{9} = \frac{5 - 1}{9} = \frac{4}{9}$

Example 1 Simplify: **a.** $\dfrac{5x}{12} + \dfrac{x}{12}$ **b.** $\dfrac{8x + 10}{9} - \dfrac{2x + 1}{9}$

Solution To add or subtract fractions with the same denominator, you add or subtract their numerators and write the result over the common denominator. Then simplify.

a. $\dfrac{5x}{12} + \dfrac{x}{12} = \dfrac{5x + x}{12}$ **b.** $\dfrac{8x + 10}{9} - \dfrac{2x + 1}{9} = \dfrac{8x + 10 - (2x + 1)}{9}$

$\qquad\qquad\quad = \dfrac{6x}{12}$ $\qquad\qquad\qquad\qquad\quad = \dfrac{8x + 10 - 2x - 1}{9}$

$\qquad\qquad\quad = \dfrac{\cancel{6} \cdot x}{\cancel{6} \cdot 2}$ $\qquad\qquad\qquad\qquad\quad = \dfrac{6x + 9}{9}$

$\qquad\qquad\quad = \dfrac{x}{2}$ $\qquad\qquad\qquad\qquad\quad = \dfrac{\cancel{3}(2x + 3)}{\cancel{3} \cdot 3}$

$\qquad\qquad\qquad\qquad\qquad\qquad\qquad\qquad\quad = \dfrac{2x + 3}{3}$

Simplify.

1. $\dfrac{x}{15} + \dfrac{4x}{15}$ **2.** $\dfrac{7x}{12} - \dfrac{5x}{12}$ **3.** $\dfrac{3}{x} - \dfrac{5}{x}$ **4.** $\dfrac{3}{5x} + \dfrac{4}{5x}$

5. $\dfrac{3x}{4} - \dfrac{x - 1}{4}$ **6.** $\dfrac{2a}{9} - \dfrac{4a + 5}{9}$ **7.** $\dfrac{y - 2}{3} - \dfrac{3y - 5}{3}$ **8.** $\dfrac{x + 5}{2} - \dfrac{7x + 1}{2}$

Example 2 Simplify: **a.** $\dfrac{2}{x + 3} + \dfrac{1}{x + 3}$ **b.** $\dfrac{3}{x - 2} + \dfrac{5}{2 - x}$

Solution **a.** $\dfrac{2}{x + 3} + \dfrac{1}{x + 3} = \dfrac{2 + 1}{x + 3}$ Add the numerators.

$\qquad\qquad\qquad\qquad\quad = \dfrac{3}{x + 3}$

b. $\dfrac{3}{x - 2} + \dfrac{5}{2 - x} = \dfrac{3}{x - 2} + \dfrac{5}{-(x - 2)}$

$\qquad\qquad\qquad\qquad\quad = \dfrac{3}{x - 2} - \dfrac{5}{x - 2}$ $\left\{\begin{array}{l}\text{Since } 2 - x = -(x - 2), \\ \text{the LCD is } x - 2.\end{array}\right.$

$\qquad\qquad\qquad\qquad\quad = \dfrac{3 - 5}{x - 2}$ Subtract the numerators.

$\qquad\qquad\qquad\qquad\quad = \dfrac{-2}{x - 2}, \text{ or } -\dfrac{2}{x - 2}$

AME _____ DATE _____

–5 Adding and Subtracting Fractions (continued)

Simplify.

9. $\dfrac{3}{x+4} + \dfrac{2}{x+4}$

10. $\dfrac{8}{y-4} - \dfrac{6}{y-4}$

11. $\dfrac{x}{x-1} + \dfrac{2}{x-1}$

12. $\dfrac{n}{n-3} - \dfrac{2n-1}{n-3}$

13. $\dfrac{4}{x-3} + \dfrac{2}{3-x}$

14. $\dfrac{7}{2x-3} - \dfrac{2}{3-2x}$

15. $\dfrac{3x}{x-y} + \dfrac{3y}{y-x}$

16. $\dfrac{4a}{a-b} + \dfrac{4b}{b-a}$

Example 3 Simplify: **a.** $\dfrac{5}{2m} - \dfrac{1}{6m^2}$ **b.** $\dfrac{a}{6} - \dfrac{4+3a}{10}$

Solution To add or subtract fractions with different denominators, first rewrite the fractions using their least common denominator.

a. $\dfrac{5}{2m} - \dfrac{1}{6m^2} = \dfrac{5 \cdot 3m}{2m \cdot 3m} - \dfrac{1}{6m^2}$ **b.** $\dfrac{a}{6} - \dfrac{4+3a}{10} = \dfrac{a \cdot 5}{6 \cdot 5} - \dfrac{(4+3a) \cdot 3}{10 \cdot 3}$

$\quad = \dfrac{15m}{6m^2} - \dfrac{1}{6m^2}$ $\qquad\qquad = \dfrac{5a - 3(4+3a)}{30}$

$\quad = \dfrac{15m-1}{6m^2}$ $\qquad\qquad\qquad = \dfrac{5a - 12 - 9a}{30}$

$\qquad\qquad\qquad\qquad\qquad\qquad\qquad = \dfrac{-4a-12}{30}$

$\qquad\qquad\qquad\qquad\qquad\qquad\qquad = \dfrac{-4(a+3)}{30}$

$\qquad\qquad\qquad\qquad\qquad\qquad\qquad = \dfrac{-2(a+3)}{15}, \text{ or } -\dfrac{2(a+3)}{15}$

Simplify.

17. $\dfrac{3}{n^2} + \dfrac{2}{n}$

18. $\dfrac{6}{x^2y} - \dfrac{4}{xy}$

19. $\dfrac{5}{2x} - \dfrac{1}{4x^2}$

20. $\dfrac{1}{3x^2} + \dfrac{5}{6x}$

21. $\dfrac{1+4x}{6} + \dfrac{1-x}{8}$

22. $\dfrac{4b+1}{8} + \dfrac{2b-3}{10}$

23. $\dfrac{a-6}{6} - \dfrac{5-a}{15}$

24. $\dfrac{1+3x}{2} - \dfrac{6x-1}{6}$

25. $\dfrac{5}{3(x+1)} - \dfrac{x}{(x+1)}$

26. $\dfrac{3x}{x-2} - \dfrac{1}{2(x-2)}$

27. $\dfrac{3}{4(x+1)} + \dfrac{x}{x+1}$

28. $\dfrac{2x}{x-1} + \dfrac{3}{2(x-1)}$

29. $\dfrac{4n-3}{15} - \dfrac{3(n-2)}{10}$

30. $\dfrac{3(a-6)}{10} - \dfrac{4(a+6)}{6}$

31. $\dfrac{3x+4}{4} - \dfrac{4x}{5} + \dfrac{x-2}{10}$

32. $\dfrac{3n}{4} + \dfrac{n+2}{3} - \dfrac{1-n}{6}$

Mixed Review Exercises

Simplify.

1. $-4^2 \cdot 3$

2. $(3 \cdot 5 - 10)^2$

3. $(-3x^2)^3$

4. $3y(y-4) + 2y(y+9)$

5. $-\dfrac{1}{5}(-30x + 20y)$

6. $(4x^2y)(3x^3y^2)(5y^2)$

7. $\left(\dfrac{6x^2y^2}{7}\right)\left(\dfrac{-14xy^2}{3}\right)$

8. $n^2(n+6) - (2n^2 - 4)n$

9. $(3 - 2 \cdot 10)^2$

6–6 Mixed Expressions

Objective: To write mixed expressions as fractions in simplest form.

Vocabulary

Mixed number The sum of an integer and a fraction. For example, $2\frac{1}{3}$.

Mixed expression The sum or difference of a polynomial and a fraction. For example, $m + \frac{3}{m}$.

Example 1   Write $2\frac{1}{3}$ as a fraction in simplest form.

Solution $2\frac{1}{3} = 2 + \frac{1}{3}$

$\qquad\quad = \frac{2}{1} + \frac{1}{3}$ Write 2 as $\frac{2}{1}$.

$\qquad\quad = \frac{6}{3} + \frac{1}{3}$ LCD = 3

$\qquad\quad = \frac{7}{3}$

Write as a fraction in simplest form

1. $3\frac{2}{3}$ 2. $2\frac{1}{8}$

3. $-3\frac{3}{5}$ 4. $-4\frac{5}{7}$

5. $5\frac{1}{6}$ 6. $6\frac{1}{5}$

7. $-2\frac{3}{4}$ 8. $-1\frac{2}{9}$

Example 2   Write each expression as a fraction in simplest form.

$\quad$ **a.** $x + \frac{2}{x}$ **b.** $3 - \frac{x-1}{x+2}$

Solution   **a.** $x + \frac{2}{x} = \frac{x}{1} + \frac{2}{x}$ Write x as $\frac{x}{1}$.

$\qquad\qquad\qquad = \frac{x^2}{x} + \frac{2}{x}$ LCD = x $\left(\frac{x}{1} = \frac{x \cdot x}{1 \cdot x} = \frac{x^2}{x}\right)$

$\qquad\qquad\qquad = \frac{x^2 + 2}{x}$

$\quad$ **b.** $3 - \frac{x-1}{x+2} = \frac{3}{1} - \frac{x-1}{x+2}$ Write 3 as $\frac{3}{1}$.

$\qquad\qquad\qquad = \frac{3(x+2)}{x+2} - \frac{x-1}{x+2}$ LCD = $x+2$ $\left(\frac{3}{1} = \frac{3(x+2)}{x+2}\right)$

$\qquad\qquad\qquad = \frac{3x + 6 - x + 1}{x+2}$

$\qquad\qquad\qquad = \frac{2x + 7}{x+2}$

Write each expression as a fraction in simplest form.

9. $6 + \frac{1}{x}$ 10. $2 + \frac{5}{a}$ 11. $3 - \frac{2}{x}$ 12. $5 - \frac{3}{n}$

13. $5a - \frac{2}{a}$ 14. $6n - \frac{4}{n}$ 15. $\frac{3}{y} + y$ 16. $4 - \frac{m}{n}$

–6 Mixed Expressions (continued)

Write each expression as a fraction in simplest form.

17. $2 + \dfrac{x}{y}$

18. $3 - \dfrac{2}{x + 1}$

19. $7 + \dfrac{y}{y - 2}$

20. $\dfrac{x}{x + 3} - 4$

21. $3x + \dfrac{x}{x - 1}$

22. $5x - \dfrac{x}{x + 2}$

23. $3y + \dfrac{y}{y - 2}$

24. $\dfrac{2n}{2n + 3} + 1$

25. $2x + \dfrac{x - 1}{x + 2}$

26. $3a + \dfrac{a - 3}{2a + 1}$

27. $3a^2 - \dfrac{a + 1}{2a + 3}$

28. $x^2 - \dfrac{2x + 1}{x + 1}$

Example 3   Write as a fraction in simplest form: $x + \dfrac{2x - 3}{x + 1} + \dfrac{5}{x + 1}$

Solution $x + \dfrac{2x - 3}{x + 1} + \dfrac{5}{x + 1} = \dfrac{x(x + 1)}{x + 1} + \dfrac{2x - 3}{x + 1} + \dfrac{5}{x + 1}$ The LCD is $x + 1$.

$= \dfrac{x^2 + x + 2x - 3 + 5}{x + 1}$ Add the numerators.

$= \dfrac{x^2 + 3x + 2}{x + 1}$

$= \dfrac{(x + 2)(x + 1)}{x + 1}$ Factor.

$= \dfrac{x + 2}{1}$ Simplify.

$= x + 2$

Write each expression as a fraction in simplest form.

29. $\dfrac{x}{x - 1} + \dfrac{x - 1}{x} - 2$

30. $x + \dfrac{2x + 1}{x - 2}$

31. $\dfrac{3}{x + 1} + \dfrac{x}{x + 1} - 1$

32. $\dfrac{x}{x + 3} + \dfrac{x}{x - 3} - 3$

33. $x - \dfrac{4}{x + 1} - \dfrac{3x - 1}{x + 1}$

34. $\dfrac{x - 1}{x} + \dfrac{3}{x - 1} + 2$

35. $\dfrac{2a}{a + 1} + \dfrac{3}{a - 1} - 1$

36. $2 - \dfrac{x}{x + 3} - \dfrac{1}{x + 3}$

37. $\dfrac{3x}{x - 1} + \dfrac{2}{x + 1} + 1$

Mixed Review Exercises

Simplify.

1. $\dfrac{3a - 3b}{3a + 12b}$

2. $\dfrac{a^2 - 7a + 10}{25 - a^2}$

3. $\dfrac{3x^2}{2y^2} + \dfrac{9xy}{8}$

4. $\dfrac{n^2 - 1}{3} + \dfrac{n + 1}{9}$

5. $\dfrac{8}{y^4} \cdot \dfrac{y^7}{2}$

6. $(-3b^3)^2$

Find the least common denominator.

7. $\dfrac{1}{3xy}, \dfrac{2}{y^2}$

8. $\dfrac{3}{a^2}, \dfrac{2}{ab}$

9. $\dfrac{1}{x - 2}, \dfrac{3}{x + 2}, \dfrac{5}{x^2 - 4}$

6–7 Polynomial Long Division

Objective: To divide polynomials.

Example 1   Divide $13x - 35 + 12x^2$ by $3x + 7$. Write the answer as a polynomial.

Solution   Rewrite $13x - 35 + 12x^2$ in order of decreasing degree of x as $12x^2 + 13x - 35$.

$$
\begin{array}{r}
4x - 5 \\
3x + 7 \overline{)\,12x^2 + 13x - 35} \\
\underline{12x^2 + 28x} \\
-15x - 35 \\
\underline{-15x - 35} \\
\end{array}
$$

$\left\{ \begin{array}{l} \text{Think:} \quad 12x^2 \div 3x = ? \\ \text{Multiply } 3x + 7 \text{ by } 4x \text{ and subtract.} \end{array} \right.$

$\left\{ \begin{array}{l} \text{Think:} \quad -15x \div 3x = ? \\ \text{Multiply } 3x + 7 \text{ by } -5 \text{ and subtract.} \end{array} \right.$

Remainder $\longrightarrow$ 0

Check: $12x^2 + 13x - 35 \stackrel{?}{=} (4x - 5)(3x + 7) + 0$ $\left\{ \begin{array}{l} \text{Multiply the divisor by the} \\ \text{quotient. Add the remainder.} \end{array} \right.$
  $12x^2 + 13x - 35 = 12x^2 + 13x - 35 \ \checkmark$

Therefore $\dfrac{12x^2 + 13x - 35}{3x + 7} = 4x - 5$.

Since the remainder is 0, both $3x + 7$ and $4x - 5$ are factors of $12x^2 + 13x - 35$.

Divide. Write the answer as a polynomial.

1. $\dfrac{x^2 + 8x + 15}{x + 5}$
2. $\dfrac{x^2 - 20 - x}{x - 5}$
3. $\dfrac{n^2 - 3n - 18}{n - 6}$
4. $\dfrac{-22 + n^2 - 9n}{n + 2}$

5. $\dfrac{n^2 + 7n + 10}{n + 5}$
6. $\dfrac{3x - 18 + x^2}{x - 3}$
7. $\dfrac{-3 - 5x + 2x^2}{2x + 1}$
8. $\dfrac{x^2 - 12x + 32}{x - 8}$

9. $\dfrac{x^2 - 48 + 8x}{x - 4}$
10. $\dfrac{x^2 - 8x - 33}{x - 11}$
11. $\dfrac{12 + n^2 - 7n}{n - 3}$
12. $\dfrac{2x + x^2 - 15}{x + 5}$

13. $\dfrac{6x^2 + x - 40}{3x + 8}$
14. $\dfrac{3x^2 - 10x - 8}{3x + 2}$
15. $\dfrac{3a^2 + 8a + 5}{3a + 5}$
16. $\dfrac{-20 + 2x^2 + 3x}{x + 4}$

Example 2   Divide: $\dfrac{2a^3 + 5}{a - 2}$.   Write the answer as a mixed expression.

Solution

$$
\begin{array}{r}
2a^2 + 4a + \ 8 \\
a - 2 \overline{)\,2a^3 + 0a^2 + 0a + \ 5} \\
\underline{2a^3 - 4a^2} \\
4a^2 + 0a \\
\underline{4a^2 - 8a} \\
8a + \ 5 \\
\underline{8a - 16} \\
21 \\
\end{array}
$$

$\left\{ \begin{array}{l} \text{Using zero coefficients, insert} \\ \text{missing terms in decreasing degree} \\ \text{of } a \text{ in } 2a^3 + 5. \text{ Then divide.} \end{array} \right.$

$\longleftarrow$ Remainder

Division ends when the remainder is either 0 or of lesser degree than the divisor.

(Check is on next page.)

–7 *Polynomial Long Division* (continued)

Check: $2a^3 + 5 \stackrel{?}{=} (2a^2 + 4a + 8)(a - 2) + 21$

$2a^3 + 5 \stackrel{?}{=} 2a^3 + 4a^2 + 8a - 4a^2 - 8a - 16 + 21$

$2a^3 + 5 \stackrel{?}{=} 2a^3 + (4a^2 - 4a^2) + (8a - 8a) - 16 + 21$

$2a^3 + 5 = 2a^3 + 5 \ \checkmark$

Therefore $\dfrac{2a^3 + 5}{a - 2} = 2a^2 + 4a + 8 + \dfrac{21}{a - 2}$.

Divide. Write the answer as a polynomial or a mixed expression.

7. $\dfrac{a^2 + 3a - 7}{a - 2}$

18. $\dfrac{k^2 - 7k + 13}{k - 3}$

. $\dfrac{6x^2 + x - 6}{3x + 2}$

20. $\dfrac{x^2 + 9}{x - 3}$

. $\dfrac{2a^2 + 5a - 10}{2a - 1}$

22. $\dfrac{5 - 2x + x^2}{x - 1}$

. $\dfrac{a^3 - 2a^2 - 3a + 4}{a + 1}$

24. $\dfrac{a^3 + 1}{a + 1}$

. $\dfrac{n^3 - 8}{n - 2}$

26. $\dfrac{2x^3 - 7x^2 + 15x - 6}{2x - 1}$

. $\dfrac{2x^3 + 9x^2 - 27}{x + 3}$

28. $\dfrac{x^3 + 4x^2 - 6}{x + 2}$

. $\dfrac{x^3 - 4x^2 + x + 6}{x - 2}$

30. $\dfrac{x^3 + 6x^2 - x - 30}{x - 2}$

. $\dfrac{2x^3 - x^2 - 5x - 2}{2x + 1}$

32. $\dfrac{x^3 - 3x^2 + 3x + 4}{x + 2}$

. $\dfrac{4x^3 - 8x^2 + 7x - 2}{2x - 1}$

34. $\dfrac{3x^3 - 7x^2 - 22x + 8}{3x - 1}$

. $\dfrac{8x^3 - 18x^2 + 27x + 8}{4x + 1}$

36. $\dfrac{2x^4 + 5x^3 - 4x^2 - 2x + 3}{x + 3}$

ixed Review Exercises

mplify.

. $\dfrac{x + 5}{3} + \dfrac{2x - 3}{3}$

2. $\dfrac{a^2}{a + 2} - \dfrac{4}{a + 2}$

3. $\dfrac{3}{x} + \dfrac{1}{2}$

. $\dfrac{3c + 1}{4c} + \dfrac{5}{2c}$

5. $\dfrac{3x + 1}{4} - \dfrac{2x - 3}{6}$

6. $\dfrac{x}{x^2 - 16} - \dfrac{1}{2x + 8}$

. $x + \dfrac{3}{x}$

8. $2 + \dfrac{n}{n - 2}$

9. $y + 2 + \dfrac{2y - 1}{y - 1}$

7 Applying Fractions

7–1 Ratios

Objective: To solve problems involving ratios.

Vocabulary

Ratio The ratio of one number to another is the quotient when the first number is divided by a second *nonzero* number.

Example 1   The ratio of 9 to 2 can be written as $9 \div 2$, $\dfrac{9}{2}$, or $9{:}2$.

Example 2   Write each ratio in simplest form.

        a. $18{:}63$ **b.** $15x{:}45x$ **c.** $\dfrac{8c^2 d}{12cd^2}$

Solution   First rewrite the ratio as a fraction if needed. Then simplify.

    a. $18{:}63 = \dfrac{18}{63} = \dfrac{2}{7}$,   **b.** $15x{:}45x = \dfrac{15x}{45x} = \dfrac{1}{3}$,   **c.** $\dfrac{8c^2 d}{12cd^2} = \dfrac{2c}{3d}$

          or $2{:}7$ or $1{:}3$

Write each ratio in simplest form.

1. $12{:}18$ **2.** $42{:}35$ **3.** $9{:}30$ **4.** $15{:}75$ **5.** $18x{:}27x$ **6.** $9y{:}48y$

7. $6a^2{:}12a$ **8.** $30x{:}6x$ **9.** $\dfrac{a^4}{(2a)^2}$ **10.** $\dfrac{25m^4}{45m^3}$ **11.** $\dfrac{56a^2 b}{14ab^3}$ **12.** $\dfrac{36rs^4}{12r^2 s^2}$

Example 3   Write each ratio in simplest form: **a.** 2 h:10 min   **b.** 8 in.:4 ft

Solution   To write the ratio of two quantities of the same kind, first express the measures in the same unit. Then write the ratio.

    a. $2 \text{ h}{:}10 \text{ min} = \dfrac{2 \text{ h}}{10 \text{ min}} = \dfrac{120 \text{ min}}{10 \text{ min}} = \dfrac{12}{1}$. or $12{:}1$

    b. $8 \text{ in.}{:}4 \text{ ft} = \dfrac{8 \text{ in.}}{4 \text{ ft}} = \dfrac{8 \text{ in.}}{48 \text{ in.}} = \dfrac{1}{6}$ or $1{:}6$

CAUTION   Using different units of measure will give an incorrect comparison.

Write each ratio in simplest form.

13. 20 sec:1 min   **14.** 6 days:2 wk   **15.** 2 ft:1 yd   **16.** 12 cm:1,2 m

17. 8 oz:2 lb   **18.** 1 day:30 h   **19.** 5 m:250 cm   **20.** 4 h:30 min

21. 1 lb:6 oz   **22.** 8 wks:1 yr   **23.** 3 kg:150 g   **24.** 6 kg:120 g

---

7–1 Ratios (continued)

Example 4   Write the ratio of wins to losses for a baseball team that played 72 games and won 45 of them.

Solution   45 wins out of 72 games tells you there were $72 - 45$ or 27 losses.

The number of wins to losses $= 45:27$, or $\dfrac{45}{27} = \dfrac{5}{3}$, or $5:3$.

Write each ratio in simplest form.

5. The student-teacher ratio in a school with 2376 students and 132 teachers.

6. The ratio of sunny to cloudy days in a year with 365 days, 275 of them sunny.

7. The ratio of two-door cars to four-door cars in a rental car fleet of 600 cars, 350 of which are four-door cars.

8. Ratio of boys to girls in a school of 1200 students if 660 students are girls.

9.   a. The ratio of men to women in an audience if 120 out of 300 are men.
     b. The ratio of women to men in part (a).

10.   a. The ratio of fiction books to nonfiction books in a library with 1100 fiction books and 1760 nonfiction books.
     b. The ratio of nonfiction books to fiction books in the library in part (a).

Find the ratio of (a) the perimeters and (b) the areas of each pair of figures.

11. A rectangle with sides 10 cm and 8 cm and one with sides 15 cm and 12 cm.

12. A rectangle with sides 12 in. and 16 in. and one with sides 15 in. and 20 in.

13. A rectangle with length 18 cm and perimeter 84 cm and one with length 15 cm and perimeter 70 cm.

14. A rectangle with length 20 in. and perimeter 100 in. and one with length 30 in. and perimeter 150 in.

15. A square with sides 75 cm and one with sides 1 m.

16. A square with sides 18 in. and one with sides 2 yd.

Mixed Review Exercises

Solve.

1. $5x = 21 - 2x$

2. $3(x - 5) + x = 5$

3. $4(3 + n) = 3(8 + 2n)$

4. $\dfrac{x + 3}{2} = -9$

5. $\dfrac{18 - 3y}{3} = 2y$

6. $-\dfrac{c}{7} = 3$

7. $(r + 3)(r - 4) = 0$

8. $2x^2 + 10x - 28 = 0$

9. $2(x - 1) = 3(x - 2)$

Simplify.

10. $\dfrac{2b + 1}{3c} + \dfrac{b}{c}$

11. $2x + \dfrac{3}{x}$

12. $\dfrac{a}{4} + \dfrac{3a + 4}{4}$

7-2 Proportions

Objective: To solve problems using proportions.

Vocabulary

Proportion An equation that states two ratios are equal is called a proportion.

For example, $2:5 = 4:10$, or $\dfrac{2}{5} = \dfrac{4}{10}$.

(Both can be read as "2 is to 5 as 4 is to 10.")

Means and extremes In the proportion, $a:b = c:d$, b and c are called the means, and a and d are called the extremes. If $\dfrac{a}{b} = \dfrac{c}{d}$, then $ad = bc$.

(The product of the means equals the product of the extremes.)

Example 1   Solve: **a.** $\dfrac{2}{x} = \dfrac{6}{4}$ **b.** $\dfrac{3}{8} = \dfrac{-6}{4a}$ **c.** $\dfrac{2}{n} = 6$

Solution   **a.** $\dfrac{2}{x} = \dfrac{6}{4}$ You can "cross-multiply" to solve a proportion.

$2 \cdot 4 = x \cdot 6$ To do this, multiply the means and the extremes.

$8 = 6x$ Then simplify.

$\dfrac{8}{6} = x$

$\dfrac{4}{3} = x$

The solution set is $\left\{\dfrac{4}{3}\right\}$.

b. $\dfrac{3}{8} = \dfrac{-6}{4a}$ **c.** $\dfrac{2}{n} = \dfrac{6}{1}$

$3 \cdot 4a = 8(-6)$ Cross-multiply. $2 \cdot 1 = 6 \cdot n$ Cross-multiply.

$12a = -48$ Simplify. $2 = 6n$ Simplify.

$a = -4$ $\dfrac{1}{3} = n$

The solution set is $\{-4\}$.   The solution set is $\left\{\dfrac{1}{3}\right\}$.

Solve.

1. $\dfrac{x}{30} = \dfrac{3}{5}$ **2.** $\dfrac{x}{24} = \dfrac{5}{6}$ **3.** $\dfrac{5}{2} = \dfrac{30}{x}$ **4.** $\dfrac{3}{4} = \dfrac{x}{8}$ **5.** $\dfrac{x}{16} = \dfrac{6}{8}$

6. $\dfrac{2}{3} = \dfrac{4}{x}$ **7.** $\dfrac{3}{4} = \dfrac{x}{32}$ **8.** $\dfrac{x}{12} = \dfrac{54}{36}$ **9.** $\dfrac{9}{2x} = \dfrac{6}{4}$ **10.** $\dfrac{4}{3n} = 2$

11. $\dfrac{3}{x} = 4$ **12.** $-8 = \dfrac{4b}{5}$ **13.** $\dfrac{10}{3k} = \dfrac{2}{5}$ **14.** $\dfrac{12t}{-7} = \dfrac{60}{14}$

15. $\dfrac{3x}{7} = -6$ **16.** $\dfrac{4r}{3} = -12$ **17.** $-2 = \dfrac{2x}{5}$ **18.** $\dfrac{91}{x} = \dfrac{7}{3}$

19. $\dfrac{7x}{45} = \dfrac{21}{9}$ **20.** $\dfrac{x}{60} = \dfrac{9}{5}$ **21.** $\dfrac{3}{2y} = \dfrac{9}{12}$ **22.** $\dfrac{14x}{35} = \dfrac{8}{5}$

7-2 Proportions (continued)

Example 2   Solve: **a.** $\dfrac{x-3}{8} = \dfrac{3}{4}$ **b.** $\dfrac{2x-1}{3} = \dfrac{4x-3}{5}$

Solution   **a.** $\dfrac{x-3}{8} = \dfrac{3}{4}$ **b.** $\dfrac{2x-1}{3} = \dfrac{4x-3}{5}$

$$4(x-3) = 3 \cdot 8 \qquad\qquad 5(2x-1) = 3(4x-3)$$
$$4x - 12 = 24 \qquad\qquad 10x - 5 = 12x - 9$$
$$4x = 36 \qquad\qquad\qquad 4 = 2x$$
$$x = 9 \qquad\qquad\qquad\quad 2 = x$$

The solution set is {9}.   The solution set is {2}.

Solve.

23. $\dfrac{x-1}{6} = \dfrac{2}{3}$

24. $\dfrac{x-2}{8} = \dfrac{3}{4}$

25. $\dfrac{3+2n}{7} = 5$

26. $3 = \dfrac{2-5y}{4}$

27. $\dfrac{x+1}{6} = \dfrac{4}{3}$

28. $\dfrac{x-4}{6} = \dfrac{3}{2}$

29. $\dfrac{x-3}{8} = \dfrac{3}{4}$

30. $\dfrac{3x+4}{4} = \dfrac{5}{2}$

31. $\dfrac{2x-1}{7} = 5$

32. $\dfrac{2n-1}{5} = 9$

33. $\dfrac{x+6}{8} = \dfrac{x-6}{9}$

34. $\dfrac{x+2}{9} = \dfrac{x-2}{3}$

35. $\dfrac{x+3}{3} = \dfrac{4x-9}{5}$

36. $\dfrac{x-6}{4} = \dfrac{x-9}{2}$

37. $\dfrac{2y-1}{3} = \dfrac{4y-3}{7}$

38. $\dfrac{3x+4}{4} = \dfrac{2x+5}{5}$

39. $\dfrac{3x-2}{4} = \dfrac{x+4}{2}$

40. $\dfrac{5x-3}{9} = \dfrac{3x+3}{7}$

Mixed Review Exercises

Find the LCD for each group of fractions.

1. $\dfrac{1}{3xy^2}, \dfrac{2}{xy}$

2. $\dfrac{w+2}{2}, \dfrac{3w-1}{9}$

3. $\dfrac{2}{9}, \dfrac{7}{12}$

4. $\dfrac{1}{3}, \dfrac{1}{4}, \dfrac{5}{24}$

5. $\dfrac{2}{x-2}, \dfrac{4}{x+2}$

6. $\dfrac{x}{2y}, \dfrac{x-1}{3}$

Simplify.

7. $\dfrac{8}{2(x+1)} + \dfrac{2}{x+1}$

8. $\dfrac{3r}{4} + \dfrac{r-1}{12}$

9. $\dfrac{3a}{8} + \dfrac{2a+1}{4}$

10. $|-6.3| - |2.7|$

11. $|-2.7| + |1.2|$

12. $6 + 2 \cdot 7$

7–3 Equations with Fractional Coefficients

Objective: To solve equations with fractional coefficients.

Example 1   Solve: **a.** $\dfrac{x}{2} + \dfrac{x}{3} = 5$ **b.** $\dfrac{2a}{5} - \dfrac{a}{4} = \dfrac{9}{20}$.

Solution   Multiply both sides of the equation by the LCD of *all* of the fractions in the equations. You will get a new equation with no fractions in it that will be easier to solve than the original equation.

a. $6\left(\dfrac{x}{2} + \dfrac{x}{3}\right) = 6(5)$ The LCD of the fractions is 6.

$$6\left(\dfrac{x}{2}\right) + 6\left(\dfrac{x}{3}\right) = 30$$
$$3x + 2x = 30 \longleftarrow \text{New equation with no fractions}$$
$$5x = 30$$
$$x = 6$$

The solution set is $\{6\}$.

b. $20\left(\dfrac{2a}{5} - \dfrac{a}{4}\right) = 20\left(\dfrac{9}{20}\right)$ The LCD of the fractions is 20.

$$20\left(\dfrac{2a}{5}\right) - 20\left(\dfrac{a}{4}\right) = 9$$
$$8a - 5a = 9 \longleftarrow \text{New equation with no fractions}$$
$$3a = 9$$
$$a = 3$$

The solution set is $\{3\}$.

Solve.

1. $\dfrac{w}{2} + \dfrac{w}{3} = \dfrac{5}{3}$

2. $\dfrac{x}{3} - \dfrac{x}{4} = \dfrac{1}{12}$

3. $\dfrac{2y}{3} + \dfrac{y}{4} = \dfrac{11}{6}$

4. $\dfrac{3a}{4} - \dfrac{4a}{3} = -\dfrac{7}{6}$

5. $\dfrac{x}{3} + \dfrac{x}{4} = 7$

6. $\dfrac{2x}{3} - \dfrac{x}{2} = 12$

7. $\dfrac{x}{4} + \dfrac{x}{5} = \dfrac{9}{5}$

8. $\dfrac{3y}{4} - \dfrac{y}{6} = \dfrac{7}{3}$

9. $\dfrac{a}{3} + \dfrac{a}{4} = -\dfrac{7}{4}$

10. $\dfrac{2x}{3} - \dfrac{5x}{9} = -1$

11. $\dfrac{a}{3} - \dfrac{a}{9} = 2$

12. $\dfrac{2a}{3} - \dfrac{3a}{2} = \dfrac{5}{6}$

13. $\dfrac{2a}{5} - \dfrac{a}{2} = \dfrac{3}{10}$

14. $\dfrac{6b}{7} - \dfrac{b}{2} = 5$

15. $\dfrac{3n}{8} + \dfrac{n}{2} = 7$

16. $\dfrac{3n}{10} + \dfrac{n}{5} = \dfrac{3}{2}$

17. $\dfrac{7m}{8} - \dfrac{m}{4} = -\dfrac{5}{2}$

18. $\dfrac{5x}{6} - \dfrac{3x}{8} = \dfrac{11}{2}$

-3 Equations with Fractional Coefficients (continued)

Example 2   Solve: **a.** $\frac{x}{3} + \frac{x-2}{4} = 0$ **b.** $3n + \frac{n}{2} = \frac{n}{3} + 19$

Solution   **a.** The LCD of the fractions is 12.   **b.** The LCD of the fractions is 6.

$$12\left(\frac{x}{3} + \frac{x-2}{4}\right) = 12(0) \qquad\quad 6\left(3n + \frac{n}{2}\right) = 6\left(\frac{n}{3} + 19\right)$$

$$12\left(\frac{x}{3}\right) + 12\left(\frac{x-2}{4}\right) = 0 \qquad 6(3n) + 6\left(\frac{n}{2}\right) = 6\left(\frac{n}{3}\right) + 6(19)$$

$$4x + 3(x-2) = 0 \qquad\qquad\qquad 18n + 3n = 2n + 114$$

$$4x + 3x - 6 = 0 \qquad\qquad\qquad\quad 21n = 2n + 114$$

$$7x - 6 = 0 \qquad\qquad\qquad\qquad 19n = 114$$

$$7x = 6 \qquad\qquad\qquad\qquad\quad n = 6$$

$$x = \frac{6}{7} \qquad\qquad\text{The solution set is } \{6\}.$$

The solution set is $\left\{\frac{6}{7}\right\}$.

Solve.

. $\frac{x}{2} - \frac{x-1}{3} = 5$

20. $\frac{x}{8} - \frac{x+3}{5} = \frac{3}{4}$

. $\frac{n}{3} - \frac{n+5}{2} = 0$

22. $\frac{x}{2} - \frac{x+3}{5} = 3$

. $\frac{x+2}{2} = \frac{2x}{3}$

24. $x + \frac{x-2}{8} = 20$

. $\frac{x+1}{4} = \frac{x-2}{3}$

26. $x + \frac{x}{2} = 7 - \frac{x}{4}$

. $\frac{x-1}{6} + \frac{x+2}{3} = 5$

28. $0 = 2m - \frac{3m+18}{6}$

. $\frac{x+1}{5} - \frac{x-1}{3} = -2$

30. $\frac{x-3}{5} + \frac{2}{3} = \frac{x+2}{15}$

. $\frac{x+1}{5} = \frac{3x-6}{10} + \frac{3}{2}$

32. $\frac{x+5}{2} - \frac{x+6}{3} = \frac{x}{4}$

. $\frac{3n-1}{7} - \frac{2n-1}{3} = -6$

34. $\frac{x+6}{6} - \frac{x}{9} = \frac{2}{3}$

lixed Review Exercises

rite each ratio in simplest form.

. 6 feet : 3 yards

2. $12x : 72x$

3. $15 : 10$

. $\frac{12m^2 n}{30mn}$

5. $\frac{36xy^2}{24x^2 y}$

6. $\frac{14a^3}{35ab^2}$

lve.

. $\frac{5}{2n} = \frac{3}{6}$

8. $\frac{x+1}{3} = \frac{5}{2}$

9. $\frac{3a+2}{4} = \frac{a+9}{3}$

. $3x - 1 = 14$

11. $|x| = 6$

12. $6x + 5 = 7x + 3$

7–4 *Fractional Equations*

Objective: To solve fractional equations.

Vocabulary

Fractional equation An equation with a variable in the denominator of one or more terms. For example, $\frac{3}{x} - \frac{1}{4} = \frac{1}{12}$. To solve a fractional equation, multiply both sides by the LCD to eliminate fractions.

CAUTION Multiplying both sides of an equation by a variable expression sometimes results in an equation that has an extra root. You must check each root of the transformed equation to see if it satisfies the original equation.

Example 1   Solve: $\frac{2}{x} + \frac{1}{4} = \frac{3}{4}$

Solution $4x\left(\frac{2}{x} + \frac{1}{4}\right) = 4x\left(\frac{3}{4}\right)$ $\left\{\begin{array}{l}\text{Multiply both sides of the equation} \\ \text{by the LCD, } 4x.\end{array}\right.$

$4x\left(\frac{2}{x}\right) + 4x\left(\frac{1}{4}\right) = 3x$ $\left\{\begin{array}{l}\text{Notice that } x \text{ cannot equal 0 because} \\ \frac{2}{0} \text{ has no meaning.}\end{array}\right.$

$\qquad\qquad\qquad 8 + x = 3x$

$\qquad\qquad\qquad\quad\ \ 8 = 2x$

$\qquad\qquad\qquad\quad\ \ 4 = x$

$\quad\quad$ *Check:* $\frac{2}{4} + \frac{1}{4} \overset{?}{=} \frac{3}{4}$ $\qquad$ $\frac{3}{4} = \frac{3}{4}\ \checkmark$ $\quad$ The solution set is $\{4\}$.

Solve and check. If the equation has no solution, write *No solution*.

1. $\frac{1}{3} + \frac{11}{x} = 4$

2. $\frac{16}{x} - \frac{3}{5} = 1$

3. $\frac{1}{2} + \frac{3}{x} = 2$

4. $\frac{1}{6} + \frac{2}{x} = \frac{5}{6}$

5. $\frac{3}{y} - \frac{1}{4} = \frac{1}{12}$

6. $\frac{1}{4} + \frac{2}{x} = \frac{3}{8}$

7. $\frac{5}{x} + \frac{3}{4} = 2$

8. $\frac{1}{x} - \frac{1}{2} = -\frac{1}{3}$

9. $\frac{7}{2a} - \frac{3}{a} = -\frac{1}{4}$

10. $\frac{3}{n} - \frac{1}{2} = \frac{6}{3n}$

11. $\frac{2}{3a} + \frac{5}{6} = \frac{3}{2a}$

12. $\frac{2}{a} + \frac{3}{2a} = \frac{7}{6}$

Example 2   Solve: $\frac{6 - x}{4 - x} = \frac{3}{5}$

Solution 1 $5(4 - x)\left[\frac{6 - x}{4 - x}\right] = 5(4 - x)\left[\frac{3}{5}\right]$ $\quad$ Multiply both sides by the LCD, $5(4 - x)$.

$\qquad\qquad\qquad 5(6 - x) = (4 - x)(3)$ $\qquad$ Notice that x cannot equal 4.

$\qquad\qquad\quad\ 30 - 5x = 12 - 3x$

$\qquad\qquad\qquad\quad\ 18 = 2x$

$\qquad\qquad\qquad\qquad 9 = x$ $\qquad\qquad\qquad$ The solution set is $\{9\}$.

–4 Fractional Equations (continued)

Solution 2 $\dfrac{6-x}{4-x} = \dfrac{3}{5}$ Solve as a proportion.

$$5(6-x) = 3(4-x)$$
$$30 - 5x = 12 - 3x$$
$$18 = 2x$$
$$9 = x \qquad \text{The solution set is } \{9\}.$$

olve.

3. $\dfrac{4-x}{6-x} = \dfrac{5}{6}$

14. $\dfrac{x+4}{x-1} = 1$

15. $\dfrac{2}{3} = \dfrac{x+5}{x+7}$

6. $3 = \dfrac{x-5}{x-3}$

17. $\dfrac{x}{x-1} = \dfrac{6}{5}$

18. $\dfrac{n}{n-2} = \dfrac{5}{7}$

9. $\dfrac{x}{x+3} = \dfrac{2}{5}$

20. $\dfrac{x}{x+5} = \dfrac{3}{2}$

21. $\dfrac{x-1}{x+3} = \dfrac{1}{2}$

2. $\dfrac{5x}{x-1} = 4$

23. $\dfrac{x+1}{3x-1} = \dfrac{1}{4}$

24. $\dfrac{x-1}{x+3} = \dfrac{3}{5}$

5. $\dfrac{8}{x+3} = \dfrac{4}{x}$

26. $\dfrac{5}{x+2} = \dfrac{3}{x}$

27. $\dfrac{2}{x+3} = \dfrac{3}{x+1}$

8. $\dfrac{2x-4}{x-2} = 3$

29. $\dfrac{a+1}{2} = \dfrac{1}{a}$

30. $\dfrac{3+x}{2x} = \dfrac{1}{x}$

1. $\dfrac{a+2}{6} = \dfrac{4}{a}$

32. $\dfrac{1}{x} + \dfrac{3x}{x-2} = 0$

33. $\dfrac{4}{x+1} - \dfrac{1}{x} = 1$

4. $\dfrac{12}{x+3} = \dfrac{2}{x-2}$

35. $\dfrac{2}{x+1} - 1 = \dfrac{1}{1-x}$

36. $\dfrac{2}{y+3} - \dfrac{1}{y-3} = 1$

7. $\dfrac{2}{x-1} + 3 = \dfrac{4x}{x-1}$

38. $\dfrac{3m+5}{6} - \dfrac{m}{2} = \dfrac{10}{m}$

39. $\dfrac{x-3}{x} + \dfrac{1}{x} = \dfrac{x+1}{x+4}$

0. $\dfrac{4}{x+1} - 1 = \dfrac{1}{x}$

41. $\dfrac{3}{1-n} + 2 = \dfrac{5}{1+n}$

42. $\dfrac{n-2}{n} - \dfrac{1}{n} = \dfrac{n-3}{n-6}$

lixed Review Exercises

lve.

1. $\dfrac{3a}{4} + \dfrac{2a}{5} = 23$

2. $\dfrac{x}{3} - \dfrac{x}{2} = 6$

3. $\dfrac{1}{5}(y-1) + \dfrac{1}{4}(y+2) = 3$

4. $\dfrac{8}{3} = \dfrac{2n}{9}$

5. $\dfrac{-6}{5t} = \dfrac{3}{10}$

6. $\dfrac{3z}{4} = \dfrac{27}{36}$

mplify.

7. $(5-3)^3$

8. $3x^2(2x^2 - 5 + 4x)$

9. $4 \cdot 3^2$

0. $(3n^2 + n) + (7 + n^2)$

11. $(4z^2)(3y^2z^2)$

12. $(2p^2q^3)^2$

7–5 Percents

Objective: To work with percents and decimals.

Vocabulary/Symbol

Percent (%) (Another way of saying hundredths, or divided by 100.)

Example 1   Write each percent as a decimal.

a. 30%   b. 2.5%   c. 420%   d. $\frac{1}{2}$%

Solution   a. $30\% = \frac{30}{100} = 0.30$ b. $2.5\% = \frac{2.5}{100} = \frac{25}{1000} = 0.025$

c. $420\% = \frac{420}{100} = 4\frac{20}{100} = 4.2$ d. $\frac{1}{2}\% = 0.5\% = \frac{0.5}{100} = \frac{5}{1000} = 0.00$

Write each percent as a decimal.

1. 20%   2. 35%   3. 60%   4. 75%   5. 4.2%

6. 1.5%   7. $\frac{1}{4}$%   8. $\frac{2}{5}$%   9. 260%   10. 140%

Example 2   Write each number as a percent: a. $\frac{1}{5}$ b. $\frac{2}{3}$ c. 3.5

Solution   a. $\frac{1}{5} = \frac{x}{100}$ b. $\frac{2}{3} = \frac{x}{100}$ c. $3.5 = \frac{x}{100}$

$5x = 100$ $3x = 200$ $\frac{3.5}{1} = \frac{x}{100}$

$x = 20$ $x = 66\frac{2}{3}$ $x = 350$

$\frac{1}{5} = \frac{20}{100} = 20\%$ $3.5 = \frac{350}{100} = 350\%$

$\frac{2}{3} = \frac{66\frac{2}{3}}{100} = 66\frac{2}{3}\%$

Write each number as a percent.

11. $\frac{3}{5}$ 12. $\frac{1}{4}$ 13. $\frac{7}{8}$ 14. 1.25   15. 2.7

16. $\frac{5}{6}$ 17. $1\frac{1}{2}$ 18. 0.075   19. 0.002   20. $\frac{9}{25}$

Example 3   25% of 160 is <u>what number?</u>

Solution 1 $\frac{25}{100} \cdot 160 = x$ **Solution 2** $0.25 \cdot 160 = x$

$\frac{4000}{100} = x$ $40 = x$

$40 = x$ 25% of 160 is <u>40</u>.   25% of 160 is <u>40</u>.

−5 Percents (continued)

olve.

1. 24% of 200 is what number?

3. 1% of 1600 is what number?

5. 32% of 300 is what number?

7. 16% of 325 is what number?

9. 7.2% of 250 is what number?

22. 5% of 120 is what number?

24. $3\frac{1}{2}$% of 400 is what number?

26. 16% of 85 is what number?

28. $8\frac{1}{2}$% of 12,000 is what number?

30. 120% of 40 is what number?

Example 4   24 is 20% of what number?

Solution
$$24 = \frac{20}{100} \cdot x$$
$$2400 = 20x$$
$$120 = x \qquad \text{24 is 20% of \underline{120}.}$$

Example 5   What percent of 60 is 45?

Solution
$$\frac{x}{100} \cdot 60 = 45$$
$$\frac{60x}{100} = 45$$
$$60x = 4500$$
$$x = 75 \qquad \underline{75\%} \text{ of 60 is 45.}$$

---

olve.

1. 12 is 20% of what number?

3. 21 is 3% of what number?

5. 18 is 15% of what number?

7. 48 is 32% of what number?

9. 4.2 is 50% of what number?

1. What percent of 25 is 16?

3. What percent of 56 is 7?

5. What percent of 90 is 225?

7. What percent of 120 is 24?

32. 15 is 40% of what number?

34. 18 is 6% of what number?

36. 27 is 75% of what number?

38. 225 is $33\frac{1}{3}$% of what number?

40. 32.4 is 25% of what number?

42. What percent of 72 is 18?

44. What percent of 150 is 60?

46. What percent of 220 is 132?

48. What percent of 36 is 45?

Mixed Review Exercises

actor completely.

1. $2ab^2 + 10b$ **2.** $2y^3 + 2y^2 - 4y$ **3.** $m^2 + 6m + 8$ **4.** $x^2 - 4y^2$

5. $4m^2 + 8mn - 12n^2$ **6.** $4a^2 + 7a - 2$ **7.** $a(a - c) + 3(a - c)$ **8.** $x^2 + 20x + 100$

7–6 Percent Problems

Objective: To solve problems involving percents.

Example 1 Find the change in price.
  a. The original price of the suit Carmen wants was $275. It is now on sale for $198.
  b. Calvin originally paid $90 for an old coin. It is now worth $145.

Solution To find the change in price, you calculate the difference between the original price and the new price.
  a. The price decreased by $275 − $198, or $77.
  b. The price increased by $145 − $90, or $55.

Example 2 The price of a salad bar increased from $3.00 to $3.45. What was the percent increase?

Solution

Step 1 The problem asks for the percent of increase.

Step 2 Let n = the percent of increase.

Step 3 $\dfrac{\text{percent of change}}{100} = \dfrac{\text{change in price}}{\text{original price}}$

Step 4
$$\frac{n}{100} = \frac{45}{300}$$
$$300n = 4500$$
$$n = 15$$

Step 5 The check is left to you. There was a 15% increase.

Complete the table.

	Item	Original price	New price	% of increase
1.	Shirt	$20.00	$22.00	?
2.	Sweater	$48.00	$60.00	?
3.	Tennis racket	$32.00	$36.00	?
4.	Movie ticket	$4.00	$5.00	?
5.	Bus ticket	$40.00	?	5%
6.	Newspaper	$.25	?	60%
7.	Books	$80.00	?	20%
8.	Magazine subscription	?	$15.00	25%
9.	Taxi fare	?	$14.00	$33\frac{1}{3}$%
10.	Airplane ticket	?	$168.00	5%

7–6 Percent Problems (continued)

Example 3 The price of a video camera decreased from $800 to $760. What was the percent decrease?

Solution

Step 1 The problem asks for the percent of decrease.

Step 2 Let x = the percent of decrease.

Step 3 Use the formula $\dfrac{\text{percent of change}}{100} = \dfrac{\text{change in price}}{\text{original price}}$.

Step 4
$$\frac{x}{100} = \frac{40}{800}$$
$$800x = 4000$$
$$x = 5$$

Step 5 The check is left to you. There was a 5% decrease.

Complete the table.

	Item	Original price	New price	% of decrease
1.	Video tape	$6.00	$4.50	?
2.	Baseball glove	$32.00	$28.00	?
3.	Skates	$60.00	$51.00	?
4.	T-shirt	$7.50	$6.00	?
5.	Watch	$80.00	?	30%
6.	Video rental	$2.00	?	25%
7.	Audio cassette	$6.00	?	20%
8.	Camera	?	$210.00	12.5%
9.	Film	?	$4.00	20%
10.	Suit	?	$136.00	15%

Mixed Review Exercises

Solve. If the equation has no solution, write *No solution*.

1. $\dfrac{2x+3}{x-3} = 5$
2. $\dfrac{a+2}{3} = \dfrac{1}{a}$
3. $\dfrac{4}{n} = \dfrac{7}{3n}$
4. $1.3x = 52$
5. $m^2 + 4m + 3 = 0$
6. $0.4x + 3.2 = 1.2x$

Simplify.

7. $(x+2)8$
8. $-3(x-2y)$
9. $8a + 5 - 3a + c$

7–7 Mixture Problems

Objective: To solve mixture problems.

Example 1 A health food store sells a mixture of raisins and roasted nuts. Raisins sell for $4.00/kg and nuts sell for $6.00/kg. How many kilograms of each should be mixed to make 40 kg of this snack worth $4.75/kg?

Solution

Step 1 The problem asks for the number of kilograms of raisins and the number of kilograms of nuts.

Step 2 Let x = the number of kilograms of raisins.
Then $40 - x$ = the number of kilograms of nuts.

	Number of kg $\times$ Price per kg $=$		Cost
Raisins	x	$4.00	$4x$
Nuts	$40 - x$	$6.00	$6(40 - x)$
Mixture	40	$4.75	190

Step 3 The value of a mixture is equal to the value of the individual ingredients.
Cost of raisins + Cost of nuts = Total cost of mixture
$4x$ $+$ $6(40 - x)$ $=$ 190

Step 4
$$4x + 6(40 - x) = 190$$
$$4x + 240 - 6x = 190$$
$$240 - 2x = 190$$
$$-2x = -50$$
$$x = 25$$
$$40 - x = 15$$

Step 5 $4(25) + 6(15) \overset{?}{=} 190$
$100 + 90 \overset{?}{=} 190$
$190 = 190 \checkmark$ 25 kg of raisins and 15 kg of nuts should be mixed.

Solve.

1. The owner of a specialty food store wants to mix cashews selling at $8.00/kg and pecans selling at $6.00/kg. How many kilograms of each should be mixed to get 12 kg of nuts worth $7.50/kg?

2. A grocer mixed 12 lb of egg noodles costing 80¢/lb with 3 lb of spinach noodles costing $1.20/lb. What will the cost per pound of the mixture be?

3. A special tea blend is made from two varieties of herbal tea, one that costs $4.00/kg and another that costs $2.00/kg. How many kilograms of each type are needed to make 20 kg of a blend worth $2.50/kg?

4. A grocer has two kinds of nuts. One costs $5/kg and another costs $4.20/kg. How many kilograms of each type of nut should be mixed in order to get 60 kg of a mixture worth $4.80/kg?

–7 **Mixture Problems** (continued)

Example 2   A chemist has 60 mL of a solution that is 70% acid. How much water should be added to make a solution that is 40% acid?

Solution

Step 1   The problem asks for the number of milliliters of water to be added.

Step 2   Let x = the number of milliliters of water to be added.

	Total amount	× % acid =	Amount of acid
Original solution	60	70%	0.70(60)
Water	x	0%	0
New solution	60 + x	40%	0.40(60 + x)

Step 3   Original amount of acid + Added acid = New amount of acid
$$0.70(60) \quad + \quad 0 \quad = \quad 0.40(60 + x)$$

Step 4
$$70(60) = 40(60 + x) \quad \left\{ \begin{array}{l} \text{Multiply both sides by} \\ \text{100 to clear decimals.} \end{array} \right.$$
$$4200 = 2400 + 40x$$
$$1800 = 40x$$
$$45 = x$$

Step 5   The check is left to you. 45 mL of water should be added.

olve.

5. A chemist has 80 mL of a solution that is 70% salt. How much water should he add to make a solution that is 40% salt?

6. If 800 mL of a juice drink is 10% grape juice, how much grape juice should be added to make a drink that is 20% grape juice?

7. How many liters of water must be added to 70 L of a 40% acid solution in order to produce a 28% acid solution?

8. How many mL of pure water must be added to 60 mL of a 20% salt solution to make a 12% salt solution?

9. A nurse has 100 mL of a solution that is 10% salt. How much sterile water must be added to make an 8% salt solution?

ixed Review Exercises

valuate.

1. 8% of 50 + 0.2% of 120

2. What percent of 60 is 18?

3. What percent of 120 is 30?

4. 12 is 25% of what number?

valuate if $a = 1$, $b = 2$, $x = 3$, and $y = 6$.

5. $|-3| + y$ 6. $\dfrac{5 - 2a}{x - b}$ 7. $\dfrac{1}{8}(6x + y)$ 8. $x^2 y$ 9. $2a + 3b$ 10. $(x - a)^2$

7–8 Work Problems

Objective: To solve work problems.

Vocabulary

Work rate The fractional part of a job done in a given unit of time. For example, if you can mow a lawn in 2 h, your work rate is $\frac{1}{2}$ job per hour. A whole job is done when the sum of the fractional parts is 1.

Example 1 Ted can paint a wall in 20 min. Vern can paint the same wall in 30 min. How long would it take them to paint the wall working together?

Solution

Step 1 The problem asks for the number of minutes needed to do the job.

Step 2 Let x = the number of minutes needed to do the job together.
Ted and Vern will each work x min.
Since Ted can do the whole job in 20 min, his work rate is $\frac{1}{20}$ job per min.
Vern's work rate is $\frac{1}{30}$ job per min.

	Work rate	× Time	= Work done
Ted	$\frac{1}{20}$	x	$\frac{x}{20}$
Vern	$\frac{1}{30}$	x	$\frac{x}{30}$

Step 3 Ted's part of the job + Vern's part of the job = whole job

$$\frac{x}{20} \quad + \quad \frac{x}{30} \quad = \quad 1$$

Step 4 $$60\left(\frac{x}{20} + \frac{x}{30}\right) = 60(1) \qquad \text{Multiply by the LCD, 60.}$$
$$3x + 2x = 60$$
$$5x = 60$$
$$x = 12$$

Step 5 The check is left to you. It would take 12 min for them to do the job together.

Solve.

1. A file clerk needs 6 h to file an average day's paperwork. It takes a trainee 12 h to do the same job. How long will it take if they work together?

2. Luis can load his truck in 24 min. It takes his brother Ramon 40 min to load the truck. How long would it take them to do the job together?

3. Ross can do a job in 8 h. Brock can do the same job in 12 h. How long would it take them working together?

4. Bernice can wallpaper a room in 4 h. Annie can wallpaper the room in 8 h. How long would it take them working together?

–8 Work Problems (continued)

Example 2   Robot A takes 6 min to weld a frame. With the help of Robot B, the job can be done in 4 min. How long would it take Robot B working alone?

Solution

Step 1   The problem asks for the amount of time it would take Robot B to weld the frame.

Step 2   Let x = the number of minutes needed for Robot B to weld the frame.

Then Robot B does $\dfrac{1}{x}$ of the job per min.

	Work rate $\times$ Time $=$ Work done		
Robot A	$\dfrac{1}{6}$	4	$\dfrac{4}{6}$
Robot B	$\dfrac{1}{x}$	4	$\dfrac{4}{x}$

Step 3   A's part of the job $+$ B's part of the job $=$ whole job.

$$\frac{4}{6} \quad + \quad \frac{4}{x} \quad = \quad 1$$

Step 4

$$6x\left(\frac{4}{6} + \frac{4}{x}\right) = 6x(1)$$
$$4x + 24 = 6x$$
$$24 = 2x$$
$$12 = x$$

Step 5   The check is left to you. It will take Robot B 12 min to weld the frame.

olve.

5. Sherry can do a job in 60 min. If her sister helps her, it takes them 36 min. How long does it take her sister alone?

6. A roofer can shingle a house in 20 h. If an apprentice helps, they can do the job in 12 h. How long does it take the apprentice alone?

7. It takes Cabin A 18 min to set the tables in the camp dining hall. If Cabin B helps them, the job can be done in 10 min. How long would it take Cabin B to set the tables by themselves?

8. One machine can print a magazine in 30 min. If a second machine works with the first machine, the magazine can be printed in 18 min. How long does it take the second machine to do the job alone?

Mixed Review Exercises

olve.

1. $\dfrac{1}{x + 2} + \dfrac{4}{x + 3} = 3$ 2. $\dfrac{6}{n} = \dfrac{12}{5}$ 3. $\dfrac{x + 2}{5} = \dfrac{x + 3}{10}$

4. $5a - 3 = 2(a + 6)$ 5. $-0.4 + k = 0.6$ 6. $-5k = 0$

NAME _____   DATE _____

7–9 Negative Exponents

Objective: To use negative exponents.

Definitions

a^{-n} If a is a nonzero real number and n is a positive integer, $a^{-n} = \dfrac{1}{a^n}$.

For example, $2^{-3} = \dfrac{1}{2^3} = \dfrac{1}{8}$. (Notice that 2^{-3} is not a negative number.)

a^0 If a is a real number not equal to zero, $a^0 = 1$. For example, $2^0 = 1$, $3^0 = 1$, $25^0 = 1$, and so on. The expression 0^0 has no meaning.

Example 1   **a.** $10^{-2} = \dfrac{1}{10^2} = \dfrac{1}{100}$ **b.** $5^{-3} = \dfrac{1}{5^3} = \dfrac{1}{125}$ **c.** $8^{-1} = \dfrac{1}{8^1} = \dfrac{1}{8}$

Simplify.

1. 3^{-1} 2. 6^{-1} 3. 4^{-3} 4. 3^{-4}

5. 5^{-2} 6. 7^{-2} 7. 2^{-5} 8. 11^{-2}

9. 9^0 10. -7^0 11. 1^{-3} 12. 5^{-1}

13. 6^{-2} 14. 5^0 15. 10^{-4} 16. 7^{-3}

All of the rules for positive exponents also hold for zero and negative exponents:

Summary of Rules for Exponents		Examples
Let m and n be any integers. Let a and b be any nonzero integers.		
1. **Products of Powers:**	$b^m b^n = b^{m+n}$	$2^2 \cdot 2^{-4} = 2^{2+(-4)} = 2^{-2} = \dfrac{1}{2^2} = \dfrac{1}{4}$
2. **Quotients of Powers:**	$b^m \div b^n = b^{m-n}$	$3^2 \div 3^5 = 3^{2-5} = 3^{-3} = \dfrac{1}{3^3} = \dfrac{1}{27}$
3. **Power of a Power:**	$(b^m)^n = b^{mn}$	$(2^2)^{-3} = 2^{-6} = \dfrac{1}{2^6} = \dfrac{1}{64}$
4. **Power of a Product:**	$(ab)^m = a^m b^m$	$(2x)^{-3} = 2^{-3} \cdot x^{-3} = \dfrac{1}{2^3} \cdot \dfrac{1}{x^3} = \dfrac{1}{8x^3}$
5. **Power of a Quotient:**	$\left(\dfrac{a}{b}\right)^m = \dfrac{a^m}{b^m}$	$\left(\dfrac{2}{3}\right)^{-2} = \dfrac{2^{-2}}{3^{-2}} = \dfrac{\frac{1}{2^2}}{\frac{1}{3^2}} = \dfrac{1}{2^2} \cdot \dfrac{3^2}{1} = \dfrac{3^2}{2^2} = \dfrac{9}{4}$

CAUTION   Remember that in $3x^2$, the exponent is applied to x but not to 3.
However, in $(3x)^2$, the exponent is applied both to 3 and to x.

–9 Negative Exponents (continued)

Example 2   Simplify. Give your answers using positive exponents.

  a. $\dfrac{3}{3^{-2}}$ b. $(x^{-1})^{-2}$ c. $(2x^{-1})^3$

Solution

  a. $\dfrac{3}{3^{-2}} = 3^{1-(-2)}$ Use the rule for quotients of powers.

   $= 3^3$

   $= 27$

  b. $(x^{-1})^{-2} = x^{(-1)(-2)}$ Use the rule for a power of a power.

   $= x^2$

  c. $(2x^{-1})^3 = 2^3 \cdot x^{(-1)(3)}$ $\begin{cases}\text{Use the rule for a power of a power and}\\\text{the rule for a power of a product.}\end{cases}$

   $= 8x^{-3}$ Use the rule for negative exponents.

   $= \dfrac{8}{x^3}$

mplify. Give your answers using positive exponents.

7. $\dfrac{2}{2^{-3}}$ **18.** $\dfrac{4^{-2}}{4^{-3}}$ **19.** $3^{-3} \cdot 3^5$ **20.** $(5^{-1})^2$ **21.** $\left(\dfrac{2^{-1}}{1}\right)^2$

. $\left(\dfrac{4^3}{4^{-2}}\right)^0$ **23.** $(5^{-1})^{-2}$ **24.** $(3^{-2})^{-1}$ **25.** $\left(\dfrac{3}{2}\right)^{-2}$ **26.** $\left(\dfrac{3^4}{3^{-2}}\right)^0$

. $\dfrac{5^{-2} \cdot 5}{5^{-1}}$ **28.** $\dfrac{3^{-4} \cdot 3^2}{3^{-2}}$ **29.** $2x^{-2}$ **30.** $3x^{-3}$ **31.** $(2x)^{-2}$

. $(3x)^{-3}$ **33.** $x^{-2}y$ **34.** $a^{-2}b^3$ **35.** $a^5 \cdot a^{-3}$ **36.** $n^3 \cdot n^{-4}$

. $(m^{-2})^3$ **38.** $(x^{-3})^2$ **39.** $(2x^{-2})^2$ **40.** $(3x^{-1})^2$ **41.** $\dfrac{y^2}{y^{-3}}$

. $\dfrac{u^{-3}}{u^7}$ **43.** $\dfrac{c^{-5}}{c^3}$ **44.** $\dfrac{d^3}{d^{-3}}$ **45.** $\dfrac{x^{-4}}{x^{-2}}$ **46.** $\dfrac{m^{-6}}{m^{-8}}$

ixed Review Exercises

mplify. Give restrictions on the variables.

. $\dfrac{24x^2y}{16xy^2}$ **2.** $\dfrac{y^2 - 7y + 10}{y^2 - 10y + 25}$ **3.** $\left(\dfrac{-2a}{b}\right)^3$

. $\dfrac{6}{5mn} - \dfrac{2}{n}$ **5.** $2 - \dfrac{4a}{a - 1}$ **6.** $\dfrac{x^2 - 4}{x^2 + 3x + 2}$

vide. Write your answer as a polynomial or as a mixed expression.

. $\dfrac{3x^2 + 10x + 3}{x + 3}$ **8.** $\dfrac{a^3 + 8}{a + 2}$ **9.** $\dfrac{32}{x^2 - 25} \div \dfrac{24}{x^2 + 10x + 25}$

7–10 *Scientific Notation*

Objective: To use scientific notation.

Vocabulary

Scientific notation A positive number in scientific notation is expressed as the product of a number greater than or equal to 1 but less than 10, and an integral power of 10. For example, 2.6×10^3 and 5.02×10^{-4} are written in scientific notation, but 0.4×10^{-5} and 10.3×10^2 are not.

Expanded notation A way of writing numbers using powers of 10 to show place value.

Example 1 Write each number in scientific notation: **a.** 34,610,000   **b.** 0.0000027

Solution   **a.** Move the decimal point left 7 places to get a number between 1 and 10.

$$34{,}610{,}000 = 3.461 \times 10{,}000{,}000 = 3.461 \times 10^7$$

b. Move the decimal point right 6 places to get a number between 1 and 10.

$$0.0000027 = \frac{2.7}{1{,}000{,}000} = \frac{2.7}{10^6} = 2.7 \times 10^{-6}$$

When a number greater than 1 is written in scientific notation, the power of 10 used is positive. When the number is less than 1, the power of 10 used is negative.

Write each number in scientific notation.

1. 27,300

2. 3,060,000

3. 25,010,000

4. 0.00305

5. 0.0000017

6. 0.000000804

Example 2 Write each number in decimal form: **a.** 3.16×10^6 **b.** 6.74×10^{-4}

Solution   **a.** Move the decimal point 6 places.   **b.** Move the decimal point 4 places.
$$3.16 \times 10^6 = 3{,}160{,}000 \qquad\qquad 6.74 \times 10^{-4} = 0.000674$$

Rewrite each number in decimal form.

7. 3.0×10^7

8. 2.27×10^8

9. 4.6×10^{11}

10. 1.8×10^5

11. 5.29×10^{-5}

12. 6.0×10^{-8}

Example 3 Write each number in expanded notation using powers of 10.
a. 7341   **b.** 0.2865   **c.** 48.09

Solution   **a.** $7341 = 7000 + 300 + 40 + 1$
$$= 7 \cdot 10^3 + 3 \cdot 10^2 + 4 \cdot 10^1 + 1 \cdot 10^0$$

b. $0.2865 = 0.2 + 0.08 + 0.006 + 0.0005$
$$= 2 \cdot 10^{-1} + 8 \cdot 10^{-2} + 6 \cdot 10^{-3} + 5 \cdot 10^{-4}$$

c. $48.09 = 40 + 8 + 0.0 + 0.09$
$$= 4 \cdot 10^1 + 8 \cdot 10^0 + 0 \cdot 10^{-1} + 9 \cdot 10^{-2}$$

-10 Scientific Notation (continued)

rite each number in expanded notation.

| . 1700 | 14. 4812 | 15. 0.143 | 16. 0.1756 | 17. 36.07 | 18. 175.1 |
| . 10,396 | 20. 0.0061 | 21. 64,000 | 22. 0.00032 | 23. 0.000015 | 24. 85,020,000 |

Example 4   Simplify. Write your answers in scientific notation.

a. $\dfrac{4.8 \times 10^6}{3.0 \times 10^2}$ b. $(1.5 \times 10^2)(8.0 \times 10^4)$ c. 0.3×10^5

Solution   a. Subtract exponents when you divide.

$$\frac{4.8 \times 10^6}{3.0 \times 10^2} = \frac{4.8}{3.0} \times \frac{10^6}{10^2}$$
$$= 1.6 \times 10^{6-2}$$
$$= 1.6 \times 10^4$$

b. Add exponents when you multiply.

$$(1.5 \times 10^2)(8.0 \times 10^4) = (1.5 \times 8.0)(10^2 \times 10^4)$$
$$= (12)(10^{2+4})$$
$$= (12)(10^6)$$
$$= 12 \times 10^6$$
$$= (1.2 \times 10) \times 10^6$$
$$= 1.2 \times 10^7$$

c. $0.3 \times 10^5 = (3 \times 10^{-1}) \times 10^5$
$$= 3 \times 10^4$$

---

nplify. Write your answers in scientific notation.

| $\dfrac{6.0 \times 10^5}{1.5 \times 10^2}$ | 26. $(4.0 \times 10^{-6})(1.6 \times 10^8)$ | 27. $(6 \times 10^6)(7 \times 10^{-2})$ |
| $(1.25 \times 10^4)(12 \times 10^3)$ | 29. $\dfrac{2 \times 10^4}{(4 \times 10^{-2})(5 \times 10)}$ | 30. $(4 \times 10^4)(7 \times 10^{-2})$ |

ixed Review Exercises

nplify. Give your answers using positive exponents.

$-[x + (-9)] + 3y + 4$	2. $(x^{-3}y^4)^2$	3. $(11 - 23) - (5 - 15)$
25:15	5. $[3 + (-5)] + 6$	6. $2t + [(-4) + (-1) + 7]$
$\left(\dfrac{c^{-2}}{2}\right)^2$	8. $-\dfrac{3}{4} + 4 + \left(-\dfrac{1}{4}\right)$	9. $\dfrac{20b^2c^3}{16bc^2}$

8 Introduction to Functions

8–1 Equations in Two Variables

Objective: To solve equations in two variables over given domains of the variables.

Vocabulary

Ordered pair A pair of numbers for which the order of the numbers is important.

Solution of an equation in two variables An ordered pair of numbers that makes the equation true.

To solve an equation To find the set of all solutions of the equation.

Symbols (a, b) (The ordered pair a, b.)

___

CAUTION 1 (x, y) is not the same as (y, x); the order is important.

___

CAUTION 2 The equation $2x + 1 = 5$ is a *one-variable equation* and has one number, $\{2\}$, for its solution. The equation $2x + y = 6$ is a *two-variable equation* and will have pairs of numbers for its solution. The numbers in a solution pair of an equation in two variables are written in the alphabetical order of the variables.

___

Example 1 State whether each ordered pair of numbers is a solution of $2x + y = 6$.

  a. $(1, 4)$ **b.** $(-1, 8)$ **c.** $(2, -2)$ **d.** $\left(\frac{5}{2}, 1\right)$

Solution Substitute each ordered pair in the equation $2x + y = 6$.

  a. $(1, 4)$ is a solution because $2(1) + 4 = 6$.

  b. $(-1, 8)$ is a solution because $2(-1) + 8 = 6$.

  c. $(2, -2)$ is *not* a solution because $2(2) + (-2) \neq 6$.

  d. $\left(\frac{5}{2}, 1\right)$ is a solution because $2\left(\frac{5}{2}\right) + 1 = 6$.

___

State whether each ordered pair is a solution of the given equation.

1. $x - y = 5$
$(6, 1), (3, -2)$

2. $2x + y = 8$
$(3, -2), (-3, -2)$

3. $x + 3y = 6$
$(3, 1)(-3, 3)$

4. $12 - y = 2x$
$(3, 6), (4, 4)$

5. $5x - 3y = 0$
$(3, 5), (-3, -5)$

6. $2x - 4y = 0$
$(2, 1), \left(1, \frac{1}{2}\right)$

7. $3a - 4b = 12$
$(4, 0), (0, 3)$

8. $2m - 3n = 6$
$(6, 2), (9, 4)$

9. $2x + 5y = 18$
$(4, 2), \left(\frac{3}{2}, 3\right)$

10. $5m - 4n = 11$
$(3, 1), \left(2, \frac{1}{4}\right)$

11. $xy = 8$
$\left(16, \frac{1}{2}\right), (-4, -2)$

12. $2xy = 4$
$\left(\frac{1}{4}, 8\right), (-2, -1)$

13. $x^2 + y^2 = 5$
$(2, -1), (3, -2)$

14. $x^2 - y^2 = 10$
$(3, -1), (1, -3)$

15. $x^2 - 2y^2 = 15$
$(5, 5), (4, 1)$

16. $2x^2 + 3y^2 = 30$
$(3, 2), (-3, 2)$

___

-1 Equations in Two Variables (continued)

Example 2   Solve $2x + 3y = 6$ for y in terms of x.

Solution
$$2x + 3y = 6$$
$$3y = 6 - 2x \qquad \text{Subtract } 2x \text{ from both sides of the equation.}$$
$$y = \frac{6 - 2x}{3} \qquad \text{Divide both sides of the equation by 3.}$$

lve each equation for y in terms of x.

. $3x + y = 6$ **18.** $2x - y = 5$ **19.** $3x + 2y = 7$

. $x + 3y = 9$ **21.** $4x + 2y = 0$ **22.** $5x + 4y = 10$

Example 3   Solve $xy + x = 4$ if x and y are whole numbers.

Solution
1. Solve the equation for y in terms of x.
$$y = \frac{4 - x}{x}$$

2. Replace x with successive whole numbers and find the corresponding values of y. If y is a whole number, you have found a solution pair.

The solutions are (1, 3), (2, 1), and (4, 0).

x	$y = \dfrac{4 - x}{x}$	Solution
0	denominator $= 0$	No
1	$\dfrac{4 - 1}{1} = 3$	(1, 3)
2	$\dfrac{4 - 2}{2} = 1$	(2, 1)
3	$\dfrac{4 - 3}{3} = \dfrac{1}{3}$	No
4	$\dfrac{4 - 4}{4} = 0$	(4, 0)

Values of x greater than 4 give negative values of y.

lve each equation if x and y are whole numbers.

. $2x + y = 4$ **24.** $3x + y = 7$ **25.** $x + 3y = 6$ **26.** $x + 2y = 5$

. $2x + 3y = 8$ **28.** $3x + y = 9$ **29.** $2x + 3y = 6$ **30.** $xy = 3$

. $xy + 1 = 7$ **32.** $xy + 2 = 9$ **33.** $xy + y = 3$ **34.** $xy - 2y = 4$

ixed Review Exercises

rite each number in scientific notation.

. 28,000,000   **2.** 0.00461   **3.** 104 million

. 0.0000325   **5.** 37,000   **6.** 6,302,000

nplify. Give answers in terms of positive exponents.

. $\dfrac{4n^2}{2n}$ **8.** $(2x)^{-3}$ **9.** $\dfrac{42x^3y^2}{14x^2y}$ **10.** $\dfrac{a^{-5}}{a^2}$

8–2 Points, Lines, and Their Graphs

Objective: To graph ordered pairs and linear equations in two variables.

Vocabulary

Plot a point Locate the graph of an ordered pair in a number plane.

Horizontal axis The horizontal number line in a number plane; the **x-axis.**

Origin The intersection of the axes on a number plane. The zero point on each axis.

Vertical axis The vertical number line in a number plane; the **y-axis.**

Graph of an ordered pair The point in a number plane associated with an ordered pair.

Abscissa The first coordinate in an ordered pair of numbers; the **x-coordinate.**

Ordinate The second coordinate in an ordered pair of numbers; the **y-coordinate.**

Coordinates of a point The abscissa and ordinate of the point, written as an ordered pair.

Coordinate axes The x- and y-axes in a number plane.

Coordinate plane A number plane; a plane in which a coordinate system has been set up.

Quadrant One of the four regions into which the coordinate axes separate a number plane.

Graph of an equation in two variables All the points that are the graphs of the solutions of the equation.

Linear equation An equation whose graph is a line.

Standard form of a linear equation The form $ax + by = c$, where a, b, and c are integers and a and b are not both zero.

Example 1 Plot each point in a number plane.

  a. $A(-3, 2)$ **b.** $B(3, -2)$ **c.** $C(-1, -3)$

Solution **a.** 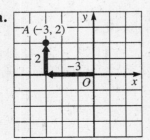   **b.** 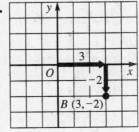   **c.**

AME _____ DATE _____

-2 Points, Lines, and Their Graphs (continued)

ot each point in a coordinate plane.

1. $A(4, 2)$ **2.** $B(6, 3)$ **3.** $C(-4, -2)$ **4.** $D(-5, -1)$

5. $E(-5, 0)$ **6.** $F(0, -5)$ **7.** $G(-3, 2)$ **8.** $H(3, -2)$

efer to the diagram at the right. Name the point(s) described.

. The point on the positive x-axis.

. The point on the negative y-axis.

. The points on the vertical line through Z.

. The points on the horizontal line through Y.

. The x-coordinate is zero.

. The y-coordinate is zero.

. The points have equal x- and y-coordinates.

. The points have opposite x- and y-coordinates.

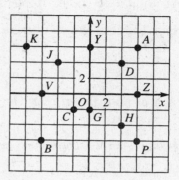

Example 2   Graph $x - 2y = 4$ in a coordinate plane.

Solution
Let $y = 0$:
$$x - 2(0) = 4$$
$$x = 4$$
Solution $(4, 0)$

Let $x = 0$:
$$0 - 2y = 4$$
$$-2y = 4$$
$$y = -2$$
Solution $(0, -2)$

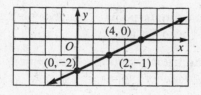

A third solution, such as $(2, -1)$ can be used as a check.

---

aph each equation. You may wish to verify your graphs on a computer or raphing calculator.

17. $x - y = 4$ **18.** $x + y = 5$ **19.** $y = 2x + 6$ **20.** $y = -2x + 2$

21. $2x + y = 4$ **22.** $x - 3y = 6$ **23.** $2x - 3y = 6$ **24.** $2x + 3y = 6$

ixed Review Exercises

te whether each ordered pair is a solution of the given equation.

1. $2x + y = 7$ **2.** $3a + 2b = 6$ **3.** $x + 3y = 11$ **4.** $2m + 3n = 7$
  $(4, -1), (-1, 9)$ $(2, -6), (2, 0)$ $(2, 3), (-3, -2)$ $(2, 1), (-1, 3)$

ve.

5. $x^2 + 5x + 6 = 0$ **6.** $-z + 9 = 3$ **7.** $2b^2 - 6b - 8 = 0$

8. $\dfrac{10 - 5y}{3} = 5$ **9.** $5x + 9 = 3x - 11$ **10.** $10 = \dfrac{2}{5}n$

8–3 Slope of a Line

Objective: To find the slope of a line.

Vocabulary

Slope If (x_1, y_1) and (x_2, y_2) are *any* two different points on a line,

$$\text{Slope} = \frac{\text{rise}}{\text{run}} = \frac{\text{difference between } y\text{-coordinates}}{\text{difference between } x\text{-coordinates}} = \frac{y_2 - y_1}{x_2 - x_1}.$$

Positive slope The slope of a line that rises from left to right is positive.

Negative slope The slope of a line that falls from left to right is negative.

Zero slope A horizontal line has slope 0.

No slope A vertical line has no slope.

Collinear points Points that lie on the same line.

Example 1 Find the slope of the line through $(-1, 3)$ and $(2, 4)$.

Solution Let $(x_1, y_1) = (-1, 3)$ and $(x_2, y_2) = (2, 4)$.

$$\text{Slope} = \frac{y_2 - y_1}{x_2 - x_1} = \frac{4 - 3}{2 - (-1)} = \frac{1}{3}$$

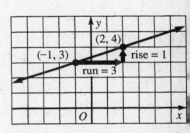

Example 2 Find the slope of the line through $(1, -3)$ and $(4, -3)$.

Solution $\text{Slope} = \frac{-3 - (-3)}{4 - 1} = \frac{0}{3} = 0$ The line has slope 0.

Example 3 Find the slope of the line through $(2, -1)$ and $(2, 5)$.

Solution $\text{Slope} = \frac{5 - (-1)}{2 - 2} = \frac{6}{0}$ (undefined)   The line has *no* slope.

Find the slope of the line through the given points.

1. $(5, -6)$, $(2, -4)$

2. $(-3, 6)$, $(-5, 4)$

3. $(0, 1)$, $(2, -2)$

4. $(1, 2)$, $(4, 6)$

5. $(2, 1)$, $(8, -2)$

6. $(-1, 5)$, $(0, 0)$

7. $(4, 3)$, $(2, 7)$

8. $(5, 2)$, $(-1, 2)$

9. $(-3, -4)$, $(1, 2)$

10. $(-5, 2)$, $(7, -6)$

11. $(1, 4)$, $(-3, 0)$

12. $(4, 4)$, $(-4, 6)$

13. $(8, -1)$, $(6, 0)$

14. $(3, -1)$, $(-2, 4)$

15. $(7, 4)$, $(7, -4)$

–3 Slope of a Line (continued)

Example 4   Find the slope of the line with the equation $2x + 3y = 6$.

Solution   1. First find any two points on the line.

If $x = 0$: $2(0) + 3y = 6$ If $y = 0$: $2x + 3(0) = 6$

$3y = 6$ $2x = 6$

$y = 2$ $x = 3$

One point: $(0,2)$ Another point: $(3,0)$

2. Now use the slope formula. Slope $= \dfrac{y_2 - y_1}{x_2 - x_1} = \dfrac{0 - 2}{3 - 0} = -\dfrac{2}{3}$

---

nd the slope of each line. If the line has no slope, say so.

6. $y = 2x - 1$ **17.** $y = 3x + 2$ **18.** $y = 4 - 2x$ **19.** $y = 6 - 3x$

0. $6x + 2y = 3$ **21.** $2x - 5y = 10$ **22.** $3x + 6y = 12$ **23.** $x - 2y = 4$

4. $y = 5$ **25.** $y + 2 = 0$ **26.** $x = 1$ **27.** $2x - 3 = 0$

Example 5   Draw a line through the point $P\,(-1, 2)$ with a slope of 3.

Solution   1. Plot point P.

2. Write the slope as $\dfrac{3}{1}$. Rise $= 3$. Run $= 1$.

3. From P, measure 1 unit to the right and 3 units up to locate a second point, T.

4. Draw the line through P and T.

---

arough the given point, draw a line with the given slope.

. $A(2, 1)$; slope 2   **29.** $B(-2, 3)$; slope -3 **30.** $C(1, -4)$; slope 4

. $D(-3, -2)$; slope $\dfrac{2}{3}$ **32.** $E(-4, 1)$: slope $-\dfrac{1}{2}$ **33.** $F(3, 0)$; slope $-\dfrac{3}{4}$

. $G(-2, -1)$; slope $\dfrac{2}{5}$ **35.** $H(-5, 2)$; slope -2 **36.** $I(2, -3)$; slope -1

---

ixed Review Exercises

lve.

. $\dfrac{x + 2}{2} + \dfrac{x}{4} = 0$ **2.** $-3 = \dfrac{9b}{4}$ **3.** $\dfrac{2 + z}{3z} = \dfrac{4}{z}$ **4.** $-3(y + 2) = 9$

aluate if $x = -2$, $y = 1$, $a = 3$, and $b = -4$.

. $\dfrac{a + 2b}{2a - b}$ **6.** $3(x + 3y)$ **7.** $\dfrac{1}{2}(3x + 4y)$ **8.** $(2a - 3b) + 5$

8–4 The Slope-Intercept Form of a Linear Equation

Objective: To use the slope-intercept form of a linear equation.

Vocabulary

y-intercept The y-coordinate of a point where a graph intersects the y-axis. Since the point is on the y-axis, its x-coordinate is 0.

Slope-intercept form of an equation The equation of a line in the form $y = mx + b$, where m is the slope and b is the y-intercept.

Parallel lines Lines in the same plane that do not intersect. Lines with the same slope and different y-intercepts are parallel.

Example 1   Find the slope and y-intercept of each line: **a.** $y = \frac{5}{2}x + 4$ **b.** $y = \frac{5}{2}x$ **c.** $y = $

Solution   Use the slope-intercept form, $y = mx + b$.

a. $y = \frac{5}{2}x + 4$

$y = \frac{5}{2}x + 4$
$\qquad \uparrow \qquad \uparrow$
$\qquad m \qquad b$

The slope is $\frac{5}{2}$ and the y-intercept is 4.

b. $y = \frac{5}{2}x$

$y = \frac{5}{2}x + 0$
$\qquad \uparrow \qquad \uparrow$
$\qquad m \qquad b$

The slope is $\frac{5}{2}$ and the y-intercept is 0.

c. $y = 4$
$y = 0x + 4$
$\qquad \uparrow \qquad \uparrow$
$\qquad m \qquad b$

The slope is 0 and the y-intercept is 4

Find the slope and the y-intercept.

1. $y = x - 3$

2. $y = 2x + 3$

3. $y = -2$

4. $y = \frac{1}{3}x + 4$

5. $y = -\frac{1}{2}x$

6. $y = -\frac{1}{3}x - 3$

7. $y = -2x + 6$

8. $y = -4x + 8$

9. $y = -x + 5$

10. $y = x - 9$

11. $y = 3x - 2$

12. $y = 3$

Example 2   Use only the slope and y-intercept to graph $y = -\frac{2}{3}x + 4$.

Solution   1. Since the y-intercept is 4, plot $(0, 4)$.

2. Since the slope $m = -\frac{2}{3} = \frac{-2}{3} = \frac{\text{rise}}{\text{run}}$, move 3 units to the right of $(0, 4)$ and 2 units down to locate a second point.

3. Draw a line through the points.

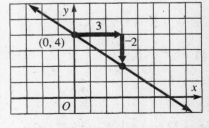

Use only the slope and y-intercept to graph each equation. You may wish to verify your graphs on a computer or a graphing calculator.

13. $y = \frac{2}{3}x - 4$

14. $y = \frac{3}{4}x - 3$

15. $y = -\frac{1}{2}x$

16. $y = -\frac{3}{4}x - 1$

17. $y = -x + 3$

18. $y = 2x + 1$

19. $y = -3$

20. $y = 5$

-4   The Slope-Intercept Form of a Linear Equation (continued)

Example 3   Use only the slope and y-intercept to graph $2x - 3y = 6$.

Solution $2x - 3y = 6$ $\begin{cases}\text{Solve for } y \text{ to transform the equation}\\ \text{into the form } y = mx + b.\end{cases}$
$-3y = -2x + 6$
$y = \dfrac{2}{3}x - 2$

1. Since $b = -2$, plot $(0, -2)$.

2. Since $m = \frac{2}{3}$, move 3 units to the right and 2 units up to locate a second point.

3. Draw a line through the points.

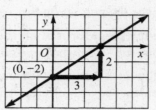

**e only the slope and y-intercept to graph each equation. You may
sh to verify your graphs on a computer or a graphing calculator.**

. $2x + y = 4$ **22.** $3x + y = 6$ **23.** $2x - y = -6$ **24.** $3x - y = 3$

. $x + 2y = -2$ **26.** $2x - 3y = 6$ **27.** $4x - 3y = 12$ **28.** $x + 4y = 4$

Example 4   Determine whether the lines with equations $4x + 5y = 20$ and $4x + 5y = 10$ are parallel.

Solution   Write each equation in slope-intercept form:

$\quad 4x + 5y = 20 \qquad\qquad\qquad 4x + 5y = 10$
$\quad\quad 5y = -4x + 20 \qquad\qquad\quad 5y = -4x + 10$
$\quad\quad\quad y = -\dfrac{4}{5}x + 4 \qquad\qquad\quad y = -\dfrac{4}{5}x + 2$

$\quad\text{slope} = -\dfrac{4}{5} \;\; y\text{-intercept} = 4 \qquad \text{slope} = -\dfrac{4}{5} \;\; y\text{-intercept} = 2$

Since both lines have the same slope and different y-intercepts, they are parallel.

termine whether the lines whose equations are given are parallel.

. $2x - y = 5$ **30.** $x - 3y = 2$ **31.** $2x - y = 6$
$\;\; 2x - y = 8$ $\quad -2x + 6y = 12$ $\quad 2y - x = 6$

. $3x - y = 2$ **33.** $\dfrac{1}{2}x - \dfrac{1}{2}y = 4$ **34.** $4x + \dfrac{1}{4}y = 2$
$\;\; -6x + 2y = 8$ $\quad 2x - 2y = 3$ $\quad 4x + 4y = 2$

ixed Review Exercises

d the slope of the line through each pair of given points.

. $(-2, 1), (-1, 2)$ **2.** $(1, 2), (3, -2)$ **3.** $(-3, 4), (-1, -2)$ **4.** $(1, 5), (2, 8)$

tor.

. $2x^2 + 7x + 6$ **6.** $2x^2 - 4x + 2$ **7.** $4y^2 - 25z^2$ **8.** $m^2 - 3mn - 10n^2$

8–5 Determining an Equation of a Line

Objective: To find an equation of a line given the slope and one point on the line, or given two points on the line.

Vocabulary

x-intercept The x-coordinate of the point where a line crosses the x-axis.

Example 1 Write an equation of a line that has slope 3 and y-intercept 2.

Solution Substitute 3 for m and 2 for b in $y = mx + b$.
The equation is $y = 3x + 2$.

Write an equation in slope-intercept form of each line described.

1. slope 2; y-intercept 3
2. slope -4; y-intercept 2

3. slope $\frac{1}{2}$; y-intercept 5
4. slope $\frac{1}{3}$; y-intercept 6

5. slope $-\frac{1}{2}$; y-intercept 4
6. slope $-\frac{1}{4}$; y-intercept 4

7. slope $\frac{2}{3}$; y-intercept -6
8. slope 3; y-intercept -7

9. slope -5; y-intercept 2
10. slope $-\frac{2}{5}$; y-intercept -1

Example 2 Write an equation of a line that has slope -2 and passes through $(5, 0)$.

Solution
1. Substitute -2 for m in $y = mx + b$
$$y = -2x + b$$

2. To find b, substitute 5 for x and 0 for y in $y = -2x + b$.
$$y = -2x + b$$
$$0 = -2(5) + b$$
$$0 = -10 + b$$
$$10 = b$$
The equation is $y = -2x + 10$.

Write an equation in slope-intercept form of each line described.

11. slope 2; passes through $(3, -1)$
12. slope 3; passes through $(-1, 2)$

13. slope -4; passes through $(2, 3)$
14. slope -2; passes through $(-3, 1)$

15. slope $\frac{2}{3}$; passes through $(0, 3)$
16. slope $-\frac{4}{3}$; passes through $(1, 0)$

17. slope $-\frac{3}{5}$; passes through $(-1, -4)$
18. slope -1; passes through $(3, 1)$

19. slope 0; passes through $\left(\frac{1}{4}, 2\right)$
20. slope 0; passes through $\left(-2, \frac{3}{8}\right)$

-5 *Determining an Equation of a Line* (continued)

Example 3 Write an equation of the line passing through the points $(-3, 2)$ and $(1, -2)$.

Solution 1. Find the slope: $\dfrac{y_2 - y_1}{x_2 - x_1} = \dfrac{-2 - 2}{1 - (-3)}$

$$= \dfrac{-4}{4} = -1$$

Substitute -1 for m in $y = mx + b$.

$$y = -x + b$$

2. Choose one of the points, say $(-3, 2)$.
Substitute -3 for x and 2 for y.

$$y = -x + b$$
$$2 = -(-3) + b$$
$$2 = 3 + b$$
$$-1 = b$$

The equation is $y = -x - 1$.

**rite an equation in slope-intercept form of the line passing through
e given points.**

. $(4, 5), (2, 1)$ **22.** $(-1, 2), (4, 7)$

. $(1, 2), (4, 4)$ **24.** $(3, 4), (4, 6)$

. $(3, 1), (5, 2)$ **26.** $(0, -2), (-3, 2)$

. $(0, -1), (-2, 3)$ **28.** $(6, 4), (2, 1)$

. $(-2, 8), (1, 2)$ **30.** $(0, 3), (-1, 0)$

. $(-1, 3), (2, 0)$ **32.** $(1, -7), (2, -1)$

rite an equation in slope-intercept form for each line described.

. y-intercept -1; x-intercept 4   **34.** y-intercept -4; x-intercept 1

. x-intercept -4; y-intercept -3 **36.** horizontal line through $(-1, -2)$

. horizontal line through $(2, 4)$ **38.** vertical line through $(-1, -2)$

ixed Review Exercises

nplify.

. $\left(\dfrac{2}{5}t^2\right)(10t^3)$ **2.** $\dfrac{1}{3}(6s^2 - 9st)$

. $(6pq^2)^2$ **4.** $(-2m^2n^3)^4$

. $2 \cdot 5 - 3^2$ **6.** $(2a^2b^3)(-3ab^2)$

. $2 \cdot (6 - 1)^2$ **8.** $(6x + 2y) - (x + y)$

8–6 Functions Defined by Tables and Graphs

Objective: To understand what a function is and to define a function by using tables and graphs.

Vocabulary

Function A correspondence between two sets, the *domain* and *range*, that assigns to each member of the domain exactly one member of the range.

Example 1   State the domain and range of the function shown by the table. Then give the correspondence as a set of ordered pairs.

High school	Northern	Central	Eastern	Western	Southern
Number of teachers	65	52	49	98	80

Solution   Domain = {Northern, Central, Eastern, Western, Southern}
Range = {49, 52, 65, 80, 98}
{(Northern, 65), (Central, 52), (Eastern, 49), (Western, 98), (Southern, 80)}

State the domain and range of each function shown by each table.
Then give each correspondence as a set of ordered pairs.

1.

Animal	Antelope	Cheetah	Greyhound	Racehorse	Rabbit
Maximum speed (mi/h)	60	70	40	50	18

2. Inventory

Item	Clock	Radio	Toaster	TV	Blender	Cookbook
Number	37	28	46	19	25	55

3. Electrical energy production

Year	1965	1970	1975	1980	1985
Billions of kilowatt-hours	1000	1500	2000	2250	2500

Example 2   Draw a bar graph for the function in the table in Example 1.

Solution   Choose the horizontal axis for the members of the domain. List the members of the range along the vertical axis. For each member of the domain, draw a vertical bar to represent the corresponding value in the range of the function. Start the scale of the bars at zero, so that the relative lengths are correct.

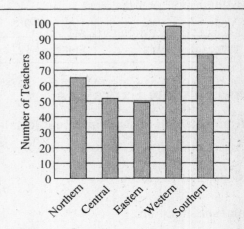

–6 Functions Defined by Tables and Graphs (continued)

–6. **Draw a bar graph for the functions shown in each table in Exercises 1–3.**

Example 3 Draw a broken-line graph for the function shown in the table.

Monthly car sales

Month	July	Aug.	Sept.	Oct.	Nov.	Dec.
Number of sales	150	170	195	205	185	200

Solution List the members of the domain along the horizontal axis. For each member of the domain plot a point to represent the corresponding value in the range of the function. Then connect the points by line segments.

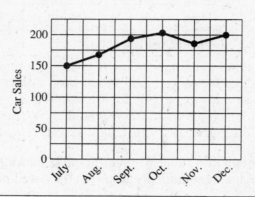

---

raw a broken-line graph for the function shown in each table.

7. Average monthly rainfall

Month	Apr.	May	June	July	Aug.	Sept.
Rainfall (mm)	60	50	85	78	40	52

Average monthly overtime

Month	July	Aug.	Sept.	Oct.	Nov.	Dec.
Hours of overtime	16	30	22	28	34	43

Average weekly pay

Year	1960	1965	1970	1975	1980	1985
Average weekly pay	$ 88	$122	$190	$289	$371	$386

8. Yearly profits

Year	Profit (in thousands)
1983	$200
1984	$215
1985	$236
1986	$270
1987	$300
1988	$350

ixed Review Exercises

rite an equation in slope-intercept form of each line described.

. passes through $(-1, 6)$ and $(-2, 10)$

2. slope -2; passes through $(-2, 1)$

. slope $\frac{1}{2}$; y-intercept -4

4. passes through $(4, 5)$ and $(5, 0)$

aph each equation.

. $y = -2x + 1$ **6.** $x + y = 5$ **7.** $y = -\frac{1}{2}x + 4$ **8.** $x = -2$

NAME _____   DATE _____

8–7 Functions Defined by Equations

Objective: To define a function by using equations.

Vocabulary

Arrow notation A notation involving an arrow used to define a function.
For example, $P: n \longrightarrow 5n - 500$.

Functional notation A notation involving an equals sign used to define a function.
For example, $P(n) = 5n - 500$.

Values of a function Members of the range of the function.

Symbols $g(2) = 6$ (Read "g of 2 equals 6" or "the value of g at 2 is 6.")

CAUTION $g(2)$ is not the product of g and 2. It names the number that g assigns to 2.

Example 1 List the range of $g: x \rightarrow x^2 - x - 6$ if the domain $D = \{-2, -1, 0, 1, 2\}$.

Solution In $x^2 - x - 6$ replace x with each member of D to find the members of the range R.

x	$x^2 - x - 6$
-2	$(-2)^2 - (-2) - 6 = 0$
-1	$(-1)^2 - (-1) - 6 = -4$
0	$(0)^2 - (0) - 6 = -6$
1	$(1)^2 - (1) - 6 = -6$
2	$(2)^2 - (2) - 6 = -4$

$R = \{0, -4, -6\}$

Note: The function g assigns -4 to both -1 and 2, and -6 to both 0 and 1.
In listing the range of g, you name -4 and -6 only once each.

Find the range of each function.

1. $g: x \rightarrow 2x + 1, D = \{-1, 0, 1\}$

2. $f: x \rightarrow 3x - 2, D = \{1, 2, 3\}$

3. $h: x \rightarrow 1 - 4x, D = \{-2, 0, 2\}$

4. $h(y) = 3y + 1, D = \{-3, 0, 1\}$

5. $G: a \rightarrow 3a - 2, D = \{-2, 0, 2\}$

6. $F(x) = 2 - 4x, D = \{-1, 0, 1\}$

7. $F(x) = 5x - 4, D = \{-1, 2, 3\}$

8. $Q(n) = 4n - 3, D = \{0, 2, 3\}$

9. $P(z) = z^2 - 2z, D = \{-1, 0, 1\}$

10. $H: b \rightarrow b^2 - b - 2, D = \{-1, 0, 2\}$

11. $g: x \rightarrow x^2 + 3x - 4, D = \{-1, 2, 4\}$

12. $f: x \rightarrow x^2 - x - 6, D = \{-2, 0, 3\}$

13. $F(x) = x^3 + x^2 + 2x, D = \{-1, 0, 1\}$

14. $N(a) = a^3 - 2a^2 + 3a, D = \{0, 2, 3\}$

-7 Functions Defined by Equations (continued)

Example 2   Given $f: x \rightarrow x^2 - x$ with the set of real numbers as the domain. Find:

   a. $f(2)$ **b.** $f(-3)$ **c.** $f(4)$

Solution   First write the equation: $f(x) = x^2 - x$

Then substitute: **a.** $f(2) = 2^2 - 2 = 4 - 2 = 2$

   b. $f(-3) = (-3)^2 - (-3) = 9 + 3 = 12$

   c. $f(4) = 4^2 - 4 = 16 - 4 = 12$

ind the values for each given function with the set of real numbers as
e domain.

5. $f(x) = 3x - 2$	**a.** $f(2)$	**b.** $f(-2)$	**c.** $f(-4)$
6. $p(x) = 4 - 2x$	**a.** $p(1)$	**b.** $p(0)$	**c.** $p(-2)$
7. $R: t \rightarrow t + 2$	**a.** $R(2)$	**b.** $R(-1)$	**c.** $R(-3)$
8. $G: n \rightarrow n - 3$	**a.** $G(0)$	**b.** $G(2)$	**c.** $G(-3)$
9. $h(a) = 2a^2 + 1$	**a.** $h(3)$	**b.** $h(-2)$	**c.** $h(0)$
10. $k(t) = 2t^2 - 3$	**a.** $k(4)$	**b.** $k(-2)$	**c.** $k(-3)$
11. $g(x) = x^2 - 1$	**a.** $g(4)$	**b.** $g(-4)$	**c.** $g(0)$
12. $h(y) = 3y^2 + 1$	**a.** $h(2)$	**b.** $h(-2)$	**c.** $h(-1)$
13. $R: y \rightarrow y^3 + 2$	**a.** $R(0)$	**b.** $R(-2)$	**c.** $R(2)$
14. $N: t \rightarrow t^3 - 8$	**a.** $N(3)$	**b.** $N(-3)$	**c.** $N(0)$
15. $f: x \rightarrow x^2 + 2x$	**a.** $f(-2)$	**b.** $f(2)$	**c.** $f(-1)$
16. $g: t \rightarrow 3t^2 - 2t$	**a.** $g(3)$	**b.** $g(1)$	**c.** $g(-1)$
17. $P(y) = y - y^2$	**a.** $P(2)$	**b.** $P(0)$	**c.** $P(-2)$

Mixed Review Exercises

implify.

1. $\dfrac{3n - 1}{2n^2} + \dfrac{2}{n}$ 2. $3\frac{1}{3} + 2\frac{3}{4} + 5\frac{2}{3} + 1\frac{1}{4}$ 3. $(-12)\left(\dfrac{x}{4}\right)$

4. $(-2)(3a + 2b - c)$ 5. $-[8 + (-3)]$ 6. $2(3m - 5)$

7. $-80\left(\dfrac{1}{4}\right)\left(\dfrac{1}{5}\right)$ 8. $\dfrac{2e^2f}{3ef^2} \cdot \dfrac{6de^2}{8ef}$ 9. $\dfrac{x^2 - 4}{x^2 + 4x + 4}$

8–8 Linear and Quadratic Functions

Objective: To graph linear and quadratic functions.

Vocabulary

Graph of a function The graph of an equation that defines a function.

Linear function A function defined by $f(x) = mx + b$. For example,
$f(x) = 2x + 3$.

Quadratic function A function defined by $f(x) = ax^2 + bx + c \ (a \neq 0)$.
For example, $f(x) = 2x^2 - x - 1$.

Parabola The graph of $f(x) = ax^2 + bx + c$, where the domain of f is the
set of real numbers and $a \neq 0$. If $a > 0$, the parabola opens upward;
if $a < 0$, the parabola opens downward.

Maximum point of a parabola The highest point on a parabola that opens
downward; the point whose y-coordinate is the *greatest value* of the
corresponding function.

Minimum point of a parabola The lowest point on a parabola that opens
upward; the point whose y-coordinate is the *least value* of the
corresponding function.

Axis of symmetry of a parabola The vertical line containing the maximum
or minimum point of the parabola. The axis of symmetry of
$y = ax^2 + bx + c \ (a \neq 0)$ is $x = -\dfrac{b}{2a}$.

Vertex of a parabola The maximum or minimum point of the parabola. The
x-coordinate of the vertex of $y = ax^2 + bx + c \ (a \neq 0)$ is $x = -\dfrac{b}{2a}$.

Example 1 Graph the function h defined by the equation $y = h(x) = -x + 2$.

Solution Find the coordinates of selected points as shown in the table below. Plot the points
and connect them with a line.

x	$-x + 2 = y$
-2	$-(-2) + 2 = 4$
-1	$-(-1) + 2 = 3$
0	$-(0) + 2 = 2$
1	$-(1) + 2 = 1$
2	$-(2) + 2 = 0$
3	$-(3) + 2 = -1$

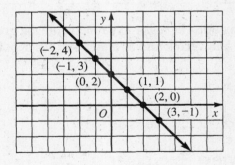

**Draw the graph of each linear function. You may wish to verify your
graphs on a computer or a graphing calculator.**

1. $g: x \rightarrow x - 2$

2. $f: x \rightarrow -x + 3$

3. $g(x) = 2 - \dfrac{1}{2}x$

4. $d(x) = -\dfrac{2}{3}x$

5. $h(x) = -4$

6. $n(x) = 5$

Study Guide, ALGEBRA, Structure and Method, Book 1

1

3–8   Linear and Quadratic Functions (continued)

Example 2   Find the coordinates of the vertex of the function $g(x) = x^2 - 2x - 3$. Then give the equation of the axis of symmetry. Use the vertex and four other points to graph the equation.

Solution

1. x-coordinate of vertex $= -\dfrac{b}{2a} = -\dfrac{-2}{2} = 1$

2. To find the y-coordinate of the vertex, substitute 1 for x.
   $y = x^2 - 2x - 3 = (1)^2 - 2(1) - 3 = 1 - 2 - 3 = -4$
   The vertex is $(1, -4)$.

3. The axis of symmetry is the line $x = 1$.

4. For values of x, select three numbers greater than 1 and three numbers less than 1 to obtain paired points with the same y-coordinate.

x	$x^2 - 2x - 3 = y$
-2	$(-2)^2 - 2(-2) - 3 = 5$
-1	$(-1)^2 - 2(-1) - 3 = 0$
0	$(0)^2 - 2(0) - 3 = -3$
1	$(1)^2 - 2(1) - 3 = -4$
2	$(2)^2 - 2(2) - 3 = -3$
3	$(3)^2 - 2(3) - 3 = 0$
4	$(4)^2 - 2(4) - 3 = 5$

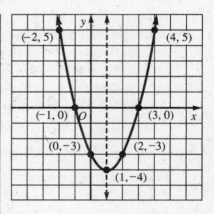

5. Plot the points. Connect them with a smooth curve.

Find the vertex and the axis of symmetry of the graph of each equation. Use the vertex and at least four other points to graph the equation.

7. $y = 2 - x^2$

8. $y = -2x^2$

9. $y = x^2 - 3x$

10. $y = -x^2 + x$

11. $y = -x^2 - x + 2$

12. $y = x^2 + 2x - 3$

Mixed Review Exercises

Find the range of each function.

1. $f(x) = 2x^2 + 3, D = \{0, 1, 2\}$

2. $m(b) = b^3 + 4, D = \{-1, 1, 2\}$

Translate each phrase into a variable expression.

3. 3 times the sum of a number and 2

4. The difference between a number and 6

5. The product of a number and 7

6. 4 less than one half of a number

8–9 Direct Variation

Objective: To use direct variation to solve problems.

Vocabulary

Direct variation A function defined by an equation of the form $y = kx$, where k is a nonzero constant. For example, $y = 5x$.

Constant of variation The nonzero constant k in a direct variation defined by $y = kx$. Also called the *constant of proportionality*.

Symbols $y = kx$ (y varies directly as x).

Example 1 Given that m varies directly as n and that $m = 75$ when $n = 25$, find the following:

a. the constant of variation   **b.** the value of m when $n = 15$

Solution Let $m = kn$.

a. Substitute $m = 75$ and $n = 25$:

$$75 = k \cdot 25$$
$$3 = k$$

b. Substitute $k = 3$ and $n = 15$: $m = 3 \cdot 15 = 45$

In Exercises 1–6, find the constant of variation.

1. y varies directly as x, and $y = 18$ when $x = 3$.

2. y varies directly as x, and $y = 52$ when $x = 13$.

3. t varies directly as s, and $t = -36$ when $s = -4$.

4. h varies directly as m, and $h = 368$ when $m = 23$.

5. y varies directly as x, and $y = 252$ when $x = 18$.

6. t varies directly as s, and $t = 490$ when $s = 14$.

Solve.

7. y varies directly as x, and $y = 300$ when $x = 5$. Find y when $x = 15$.

8. y varies directly as x, and $y = 10$ when $x = 2$. Find y when $x = 9$.

9. h varies directly as a, and $a = 20$ when $h = 4$. Find a when $h = 3$.

10. h varies directly as a, and $a = 24$ when $h = 8$. Find a when $h = 4$.

11. y varies directly as x, and $y = 240$ when $x = 25$. Find y when $x = 40$.

12. h varies directly as a, and $a = 6$ when $h = 15$. Find a when $h = 5$.

−9 **Direct Variation** (continued)

Example 2   The amount of interest earned on savings is directly proportional to the amount of money saved. If $26 interest is earned on $325, how much interest will be earned on $900 in the same period of time?

Solution 1

Step 1   The problem asks for the interest earned on $900 if the interest on $325 is $26.

Step 2   Let i, in dollars, be the interest on d dollars.

$$i_1 = 26 \qquad i_2 = \frac{?}{}$$
$$d_1 = 325 \qquad d_2 = 900$$

Step 3   An equation can be written in the form $\dfrac{i_1}{d_1} = \dfrac{i_2}{d_2}$.

$$\frac{26}{325} = \frac{i_2}{900}$$

Step 4

$$26(900) = 325i_2$$
$$23{,}400 = 325i_2$$
$$72 = i_2$$

Step 5   The check is left for you. The interest earned on $900 will be $72.

Solution 2   To solve Example 2 by the method shown in Example 1, first write the equation $i = kd$. Then solve for the constant of variation, k, by using the fact that $i = 26$ when $d = 325$. Use the value of k to find the value of i when $d = 900$. You may wish to complete the problem this way.

Solve.

3. An employee's wages are directly proportional to the time worked. If an employee earns $120 for 8 h, how much will the employee earn for 20 h?

4. A certain car used 21 gal of gasoline in 7 h. If the rate of gasoline used is constant, how much gasoline will the car use on a 6-hour trip?

5. The distance traveled by a bus at a constant speed varies with the length of time it travels. If a bus travels 192 mi in 4 h, how far will it travel in 9 h?

6. The number of words typed is directly proportional to the time spent typing. If a typist can type 325 words in 5 min, how long will it take the typist to type a 1040-word report?

Mixed Review Exercises

Multiply.

1. $(2x - 3)(3x - 1)$
2. $(3x - 2)(x^2 + x - 3)$
3. $-2x(3 - 5x)$
4. $(2x + 5)(2x - 5)$
5. $(t - 2)(3t + 5)$
6. $(5y - 3)(2y + 3)$

Study Guide, ALGEBRA, Structure and Method, Book 1

8–10 Inverse Variation

Objective: To use inverse variation to solve problems.

Vocabulary

Inverse variation A function defined by an equation of the form $xy = k$, where k is a nonzero constant. For example, $xy = 6$.

Hyperbola The graph of $xy = k$ for any nonzero value of k.

Example 1   Graph the equation $xy = -1$.

Solution

x	y
-4	$\frac{1}{4}$
-2	$\frac{1}{2}$
-1	1
$-\frac{1}{4}$	4

x	y
$\frac{1}{4}$	-4
$\frac{1}{2}$	-2
1	-1
4	$-\frac{1}{4}$

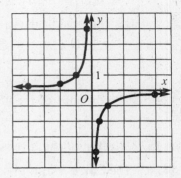

Graph each equation if the domain and the range are both the set of real numbers. You may wish to verify your graphs on a computer or graphing calculator.

1. $xy = 8$ 2. $xy = 16$ 3. $xy = -4$ 4. $xy = -6$

5. $x = \dfrac{4}{y}$ 6. $y = \dfrac{6}{x}$ 7. $\dfrac{x}{3} = \dfrac{-3}{y}$ 8. $\dfrac{x}{2} = \dfrac{6}{y}$

Example 2 (x_1, y_1) and (x_2, y_2) are ordered pairs of the same inverse variation. Find the missing value: $x_1 = 2$, $y_1 = 28$, $x_2 = 4$, $y_2 = \underline{\ ?\ }$.

Solution   An inverse variation $xy = k$ can also be expressed as $x_1y_1 = x_2y_2$.

$2 \cdot 28 = 4 \cdot y_2$ Replace x_1 with 2, y_1 with 28, and x_2 with 4.

$56 = 4y_2$ Solve the equation.

$14 = y_2$, or $y_2 = 14$.

(x_1, y_1) **and** (x_2, y_2) **are ordered pairs of the same inverse variation. Find the missing value.**

9. $x_1 = 6$, $y_1 = 5$, $x_2 = 2$, $y_2 = \underline{\ ?\ }$ 10. $x_1 = 8$, $y_1 = 24$, $x_2 = \underline{\ ?\ }$, $y_2 = 48$

11. $x_1 = 5$, $y_1 = 8$, $x_2 = 10$, $y_2 = \underline{\ ?\ }$ 12. $x_1 = 6$, $y_1 = \underline{\ ?\ }$, $x_2 = 9$, $y_2 = 8$

13. $x_1 = \underline{\ ?\ }$, $y_1 = 20$, $x_2 = 8$, $y_2 = 5$ 14. $x_1 = 8$, $y_1 = 9$, $x_2 = \underline{\ ?\ }$, $y_2 = 18$

3–10 Inverse Variation (continued)

Example 3 If a 12 g mass is 60 cm from the fulcrum of a lever, how far from the fulcrum is a 45 g mass that balances the 12 g mass?

Solution A *lever* is a bar pivoted at a point called the *fulcrum*. If masses m_1 and m_2 are placed at distances d_1 and d_2 from the fulcrum, and the bar is balanced, then $m_1 d_1 = m_2 d_2$.

Let $m_1 = 12$, $d_1 = 60$, and $m_2 = 45$, $d_2 = \underline{\ ?\ }$.

Use $m_1 d_1 = m_2 d_2$.

$12 \cdot 60 = 45 \cdot d_2$.

$720 = 45 d_2$

$16 = d_2$

The distance of the 45 g mass from the fulcrum is 16 cm.

In Exercises 15–22, refer to the lever at balance in Example 3. Find the missing value.

15. $m_1 = 12$, $m_2 = 8$, $d_1 = 45$, $d_2 = \underline{\ ?\ }$ **16.** $m_1 = 60$, $m_2 = \underline{\ ?\ }$, $d_1 = 8$, $d_2 = 12$

17. $m_1 = 24$, $m_2 = 8$, $d_1 = \underline{\ ?\ }$, $d_2 = 18$ **18.** $m_1 = \underline{\ ?\ }$, $m_2 = 40$, $d_1 = 5$, $d_2 = 7$

19. $m_1 = 12$, $m_2 = 9$, $d_1 = \underline{\ ?\ }$, $d_2 = 40$ **20.** $m_1 = 108$, $m_2 = 60$, $d_1 = \underline{\ ?\ }$, $d_2 = 9$

Solve.

21. Sarah weighs 105 lb and Wyatt weighs 140 lb. If Sarah sits 8 ft from the seesaw support, how far from the support must Wyatt sit to balance the seesaw?

22. Yoko weighs 120 lb and Lars weighs 180 lb. If Yoko sits 6 ft from the seesaw support, how far from the support must Lars sit to balance the seesaw?

Mixed Review Exercises

Show that the lines whose equations are given are parallel.

1. $x + 2y = 3$
$x + 2y = 5$

2. $2x + 6y = 7$
$x + 3y = 1$

3. $x - y = 3$
$y - x = 3$

4. $-6x + 9y = 2$
$2x - 3y = 6$

Find the constant of variation.

5. t varies directly as s, and $t = 12$ when $s = -3$.

6. y varies directly as x, and $y = 8$ when $x = 32$.

7. m varies directly as n, and $m = 27$ when $n = 3$.

9 Systems of Linear Equations

9–1 The Graphing Method

Objective: To use graphs to solve systems of linear equations.

Vocabulary

System of equations Two or more equations in the same variables. Also called a *system of simultaneous equations*.

To solve a system of equations To find all ordered pairs (x, y) that make *both* equations true.

Solution of a system of equations An ordered pair that satisfies both equations at the same time.

Coincide Two lines coincide if their graphs are the same. The equations are equivalent.

Example 1   Solve the system by graphing:

$$2x - y = 1$$
$$x + y = 5$$

Solution   Graph $2x - y = 1$ and $x + y = 5$ in the same coordinate plane. The only point on *both* lines is the *intersection point* $(2, 3)$. The only solution of *both* equations is $(2, 3)$.

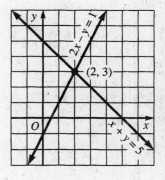

Check: You can check that $(2, 3)$ is a solution of the system by substituting $x = 2$ and $y = 3$ in both equations.

$$2x - y = 1 \qquad x + y = 5$$
$$2(2) - 3 = 1 \qquad 2 + 3 = 5$$

The system has the solution $(2, 3)$.

Solve each system by the graphing method.

1. $x + y = 6$
$x - y = 2$

2. $x + y = 5$
$x - y = -3$

3. $x + y = 9$
$x - y = 3$

4. $y = x + 2$
$y = 2x - 1$

5. $2x - y = 0$
$x + y = 3$

6. $2x + y = 1$
$x + y = 3$

7. $2x + y = 5$
$x - y = 4$

8. $x + 2y = 5$
$x - y = -1$

9. $x - y = 4$
$2x + y = 2$

10. $-2x + y = -1$
$2x + y = 7$

11. $y - 2x = -5$
$y - x = -3$

12. $2y - x = 2$
$y + x = 4$

-1 The Graphing Method (continued)

Example 2   Solve the system by graphing: $x - 2y = 4$
                                                  $x - 2y = -2$

Solution   When you graph the equations in the same coordinate plane, you see that the lines have the same slope but different y-intercepts. The graphs are parallel lines. Since the lines do not intersect, there is no point that represents a solution of both equations.

The system has *no solution*.

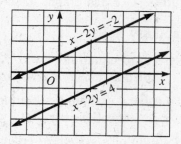

Example 3   Solve the system by graphing: $x + y = 3$
                                                  $2x + 2y = 6$

Solution   When you graph the equations in the same coordinate plane, you see that the graphs coincide. The equations are equivalent. Every point on the line represents a solution of both equations.

The system has *infinitely many solutions*.

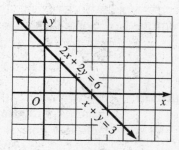

...ve each system by the graphing method.

$3x - y = 8$
$x + y = 4$

14. $2x + 3y = 5$
$y = x$

15. $2x - 3y = 4$
$2x - y = 0$

$3x + 3y = 9$
$x + y = 3$

17. $x + 2y = -4$
$x + 2y = 8$

18. $3x + y = 6$
$2x - y = -1$

$x - y = -6$
$x - y = 2$

20. $y - x = -3$
$y - 2x = -5$

21. $2x + y = 5$
$2x + y = -1$

$2x - y = 7$
$x + 2y = 11$

23. $4x + y = -14$
$3x = y$

24. $x - y = 4$
$2x - 2y = 8$

xed Review Exercises

...plify. Give your answers using positive exponents.

$\dfrac{16a^2b}{8ab^2}$

2. $(a^{-2}b^3)^3$

3. $\dfrac{15m^5n}{25m^2n^3}$

$(x^3y^2)^{-2}$

5. x^4y^{-3}

6. $\dfrac{x^3y^2}{x^{-2}y}$

9–2 The Substitution Method

Objective: To use the substitution method to solve systems of linear equations.

Example 1   Solve by the substitution method: $x + y = 9$
$2x + 3y = 20$

Solution

1. Solve the first equation for y. $x + y = 9$
$y = 9 - x$

2. Substitute this expression for y in the other equation, and solve for x.

$2x + 3(9 - x) = 20$
$2x + 27 - 3x = 20$
$-x + 27 = 20$
$-x = -7$
$x = 7$

3. Substitute the value for x in the equation in Step 1, and solve for y.

$y = 9 - x$
$y = 9 - 7$
$y = 2$

4. Check $x = 7$ and $y = 2$ in *both* equations.

$x + y = 9$ $2x + 3y = 20$
$7 + 2 \overset{?}{=} 9$ $2(7) + 3(2) \overset{?}{=} 20$
$9 = 9 \checkmark$ $14 + 6 \overset{?}{=} 20$
$20 = 20 \checkmark$

The solution is $(7, 2)$.

Solve by the substitution method.

1. $y = 3x$
    $x + y = 12$

2. $y = 2x$
    $5x - y = 12$

3. $a = 4b$
    $a - b = 9$

4. $m = 5n$
    $3m - 2n = 26$

5. $y = x - 1$
    $2x + y = 5$

6. $y = 4x - 1$
    $x + y = 4$

7. $x + y = 3$
    $2x - y = 6$

8. $x - y = 2$
    $x - 2y = -1$

9. $3x - y = -9$
    $4x + y = -5$

10. $2x + y = 1$
    $3x + 2y = 3$

11. $3x + y = 7$
    $2x - 5y = -1$

12. $x - 3y = -5$
    $2x - 5y = -9$

13. $4x - 2y = 5$
    $x - 4y = 3$

14. $2x + y = 3$
    $3x + 2y = 5$

15. $3y - x = -8$
    $5y + 2x = -6$

16. $3x + y = 2$
    $2x + 3y = -8$

17. $x + 2y = 7$
    $2x - y = 4$

18. $x - 3y = 2$
    $x = -y - 6$

19. $x - 5 = y$
    $5x + 2y = 4$

20. $y - 3 = -2x$
    $3x - 2y = -20$

21. $x + 8 = 2y$
    $4x + y = 13$

22. $3u + v = 8$
    $\dfrac{u}{4} - \dfrac{v}{2} = 3$

23. $2x - y = 2$
    $x = \dfrac{2}{3}y$

24. $5x - 4y = -10$
    $x = \dfrac{3}{5}y$

–2 The Substitution Method (continued)

Example 2   Solve by the substitution method: $2x - 6y = 8$
$x - 3y = 10$

Solution
$$x - 3y = 10$$
$$x = 10 + 3y$$

$2x - 6y = 8$
$2(10 + 3y) - 6y = 8$
$20 + 6y - 6y = 8$
$20 = 8 \leftarrow$ False

{ The *false statement* indicates that there is *no* ordered pair (x, y) that satisfies both equations. (If you graph the equations, you'll see that *the lines are parallel.*)

The system has *no solution.*

Example 3   Solve by the substitution method: $\dfrac{y}{3} = 3 - x$
$3x + y = 9$

Solution $\dfrac{y}{3} = 3 - x$ Multiply both sides by 3 to solve for y.
$y = 9 - 3x$

$3x + y = 9$
$3x + (9 - 3x) = 9$
$3x + 9 - 3x = 9$
$9 = 9 \leftarrow$ True

{ The *true statement* indicates that *every* ordered pair (x, y) that satisfies one of the equations also satisfies the other. (If you graph the equations, you'll see that *the lines coincide.*)

The system has *infinitely many solutions.*

---

olve by the substitution method.

5. $x - 3y = -2$
   $y = 2x - 1$

26. $x + 2y = 7$
   $2x + 4y = 8$

27. $y = 2x - 3$
   $2y = -3x + 8$

8. $\dfrac{x}{2} = 3 - y$
   $x + 2y = 6$

29. $9x - 5y = 105$
   $\dfrac{1}{4}x - \dfrac{2}{5}y = -1$

30. $\dfrac{x}{3} = 2 + y$
   $3x - 9y = -4$

lixed Review Exercises

/rite an equation in slope-intercept form for each line described.

1. slope $\dfrac{1}{2}$, passes through $(-2, 4)$

2. slope $\dfrac{2}{3}$, passes through $(3, -3)$

3. slope 3, y-intercept 2

4. passes through $(2, 7)$ and $(0, -3)$

5. passes through $(2, -4)$ and $(-1, 1)$

6. slope 0, y-intercept -3

9–3 Solving Problems with Two Variables

Objective: To use systems of linear equations in two variables to solve problems.

Example 1   Joel has 14 coins, all dimes and quarters, worth $2.60. How many dimes and quarters does Joel have?

Solution

Step 1   The problem asks for the number of dimes and the number of quarters.

Step 2   Let d = the number of dimes and q = the number of quarters. Make a chart.

	Number	× Value per coin	= Total value
Dimes	d	10	$10d$
Quarters	q	25	$25q$

Step 3   The two facts not recorded in the chart are the total number of coins, 14, and the total value, $2.60. Use these facts to write a system of equations.

$$d + q = 14$$
$$10d + 25q = 260$$

Step 4 $ d = 14 - q$ Find d in terms of q.
$ 10(14 - q) + 25q = 260$ Substitute.
$ 140 - 10q + 25q = 260$
$ 15q = 120$
$ q = 8$

$d = 14 - q$
$d = 14 - 8$
$d = 6$

Step 5   The check is left for you. Joel has 6 dimes and 8 quarters.

Solve, using two equations in two variables.

1. Rod has 40 coins, all dimes and quarters, worth $7.60. How many dimes and how many quarters does he have?

2. Gayle has 36 coins, all nickels and dimes, worth $2.40. How many dimes does she have?

3. Leo has $4.80 in dimes and quarters. He has 6 more dimes than quarters. How many quarters does he have?

4. Nancy and Kerry have the same number of coins. Nancy has only dimes and Kerry has only quarters. If Kerry has $3.00 more than Nancy, how much does she have?

5. Ben has $3.40 in nickels and dimes. He has 4 more dimes than nickels. How many dimes does he have?

–3   Solving Problems with Two Variables *(continued)*

Example 2   Connie has \$4000 invested in stocks and bonds. The stocks pay 6% interest and the bonds pay 8% interest. If her annual income from the stocks and bonds is \$270, how much is invested in stocks?

Solution

Step 1   The problem asks for the amount invested in stocks.

Step 2   Let s = amount invested in stocks and b = amount invested in bonds.

	Principal $\times$ Rate $=$ Interest		
Stocks	s	0.06	$0.06s$
Bonds	b	0.08	$0.08b$
Total	4000		270

Step 3 $s + b = 4000$ The total amount invested is \$4000.
$0.06s + 0.08b = 270$ The total amount of interest earned is \$270.
$s = 4000 - b$ Find s in terms of b.

Step 4 $0.06(4000 - b) + 0.08b = 270$ Substitute.
$6(4000 - b) + 8b = 27{,}000$ $\{$ Multiply each side of the equation
$24{,}000 - 6b + 8b = 27{,}000$ $\}$ by 100 to eliminate decimals.
$24{,}000 + 2b = 27{,}000$
$2b = 3000$
$b = 1500$ $s = 4000 - b$ or $s = 2500$

Step 5   The check is left for you. Connie has \$2500 invested in stocks.

---

olve, using two equations in two variables.

5. Sam invests \$6000 in treasury notes and bonds. The notes pay 8% annual interest and the bonds pay 10% annual interest. If the annual income is \$550, how much is invested in bonds?

7. Kathleen has \$8000 invested in stocks and bonds. The stocks pay her 6% annual interest and the bonds pay 9% interest. If her annual income from the stocks and bonds is \$630, how much is invested in stocks?

8. Marty invested \$7000 in treasury notes and stocks. The stocks paid 7% and the notes paid 8%, giving an annual income of \$535. How much is invested in treasury notes?

Mixed Review Exercises

Solve.

1. $\frac{1}{3}x + 3 = 1$ **2.** $\frac{1}{2}y = 3\frac{1}{2}$ **3.** $\frac{x+3}{2} = 6$

4. $2(a+1) = 8 - 4(a-6)$ **5.** $-9 = n + 4$ **6.** $3x + 15 = x + 5$

9–4 The Addition-or-Subtraction Method

Objective: To use addition or subtraction to solve systems of linear equations in two variables.

Vocabulary

Addition-or-subtraction method A method to solve systems of equations. You can use the addition-or-subtraction method whenever two equations have the same or opposite coefficients for one of their terms.

Example 1   **(The Addition Method)**

Solve: $4x - y = 7$
$2x + y = 5$

Solution

1. Add similar terms of the two equations.

$4x - y = 7$
$\underline{2x + y = 5}$ $\left\{ \begin{array}{l}\text{The } y\text{-terms} \\ \text{are eliminated.}\end{array}\right.$
$6x \quad\;\; = 12$

2. Solve the resulting equation. $x = 2$

3. Substitute 2 for x in either of the original equations to find y.

$2x + y = 5$
$2(2) + y = 5$
$y = 1$

4. Check $x = 2$ and $y = 1$ in *both* original equations.

$4x - y = 7 \qquad\qquad 2x + y = 5$
$4(2) - 1 \overset{?}{=} 7 \qquad 2(2) + 1 \overset{?}{=} 5$
$7 = 7 \qquad\qquad\quad 5 = 5$

The solution is (2, 1).

Example 2   **(The Subtraction Method)**

Solve: $5c + 3d = 14$
$5c - d = 22$

Solution

1. Subtract similar terms of the two equations.

$5c + 3d = 14$
$\underline{5c -\; d = 22}$ $\left\{ \begin{array}{l}\text{The } c\text{-terms} \\ \text{are eliminated.}\end{array}\right.$
$4d = -8$

2. Solve the resulting equation. $d = -2$

3. Substitute -2 for d in either of the original equations to find c.

$5c + 3(-2) = 14$
$5c - 6 = 14$
$5c = 20$
$c = 4$

4. The check in both equations is left for you.

The solution is (4, −2).

9–4 The Addition-or-Subtraction Method (continued)

Solve by the addition-or-subtraction method.

1. $x + y = 6$
$x - y = 2$

2. $m + n = 12$
$m - n = 6$

3. $2x + y = 3$
$x - y = 3$

4. $2x + y = 5$
$x + y = 4$

5. $3m - 2n = 11$
$5m + 2n = 13$

6. $12m + 3n = 0$
$5m + 3n = 7$

7. $6x - 7y = 14$
$-6x + 3y = -6$

8. $4a - 5b = 10$
$2a - 5b = 0$

9. $2c + 3d = 3$
$2c + d = -3$

10. $4x - 3y = -10$
$2x + 3y = 4$

11. $2x - y = 7$
$3x + y = 8$

12. $6x - 5y = 1$
$2x - 5y = 17$

13. $9x + 2y = -22$
$9x - 10y = 2$

14. $5m + 12n = -1$
$8m + 12n = 20$

15. $3a + 2c = 30$
$5a - 2c = 2$

16. $3m + 4n = 7$
$-3m + 9n = 6$

17. $4x - 2y = -8$
$4x + 5y = 6$

18. $6a - 5b = 2$
$4a + 5b = -32$

19. $7x - 11y = -1$
$13x + 11y = 61$

20. $\frac{1}{2}x + \frac{1}{3}y = 2$
$\frac{3}{2}x - \frac{1}{3}y = 2$

21. $\frac{3}{4}x - \frac{1}{6}y = -7$
$\frac{5}{4}x - \frac{1}{6}y = -11$

Solve by either the substitution or the addition-or-subtraction method.

22. $a = 4b$
$a + 2b = -6$

23. $x - 5y = 3$
$2x + y = 6$

24. $3x - 8y = 10$
$2x + 8y = -20$

25. $3(a - 2b) = 6$
$2(a + 3b) = -6$

26. $n = 6m - 2$
$\frac{1}{2}n - m = -1$

27. $\frac{1}{3}a - \frac{2}{3}b = -2$
$a + b - 12 = 0$

28. $y = \frac{2}{3}x$
$2x + 3y = -24$

29. $\frac{a}{3} - \frac{b}{3} = 2$
$2a + b = 3$

30. $2n - 11 = \frac{m}{4}$
$n = \frac{m}{-3}$

Mixed Review Exercises

Simplify.

1. $6x^3 + 4x^2 - x + 5x^2$

2. $2 \cdot 3^2$

3. $(2 \cdot 10^3) + (3 \cdot 10^2) + (5 \cdot 10)$

4. $-3[2n - (n + 1)]$

5. $(8x^3y^2)\left(\frac{3}{4}x^2y\right)$

6. $(2a^5)^2$

7. $(-2ab^2)^3$

8. $2x[3x + 2(4 - x)]$

9. $(4ab)(-2ab^2)(5a^2b^3)$

10. $\left(-\frac{1}{12}\right)(60)\left(\frac{1}{5}\right)$

11. $\dfrac{-6}{\frac{1}{2}}$

12. $\frac{1}{5}(-45m + 30n)$

9–5 Multiplication with the Addition-or-Subtraction Method

Objective: To use multiplication with the addition-or-subtraction method to solve systems of linear equations.

Example 1 Solve: $3x - y = 9$
$2x + 5y = -11$

Solution 1. Multiply both sides of the first equation by 5 so that the y-terms are opposites.

$$5(3x - y) = 5(9) \rightarrow 15x - 5y = 45$$
$$2x + 5y = -11 \rightarrow \underline{2x + 5y = -11}$$

2. Add similar terms. $17x \qquad = 34$

3. Solve the resulting equation. $x = 2$

4. Substitute 2 for x in either original equation to find the value of y.

$$3(2) - y = 9$$
$$6 - y = 9$$
$$-y = 3$$
$$y = -3$$

5. The check is left for you.

The solution is $(2, -3)$.

CAUTION Check your solution in the original equations as a transformed equation could contain an error.

Solve each system by using multiplication with the addition-or-subtraction method.

1. $2x + y = 7$
$3x - 4y = 5$

2. $3a + 5b = 1$
$a + 2b = 0$

3. $2x - y = 8$
$x - 4y = -3$

4. $m + 2n = 9$
$3m - 5n = 5$

5. $a - 2b = 1$
$3a + b = -4$

6. $3x - 2y = -1$
$x + y = 3$

7. $5x - y = -4$
$4x - 3y = -1$

8. $2m + 3n = 6$
$m + 2n = 5$

9. $2x - y = 8$
$x - 8y = 4$

10. $x + 3y = -2$
$4x + 7y = 7$

11. $x + 3y = 5$
$3x + 2y = -6$

12. $5x - 2y = -3$
$x + 3y = -4$

13. $3x - 2y = 5$
$x - 4y = -5$

14. $5x - y = 14$
$4x - 3y = 20$

15. $3x + 2y = 2$
$-7x + y = -16$

Example 2 Solve: $3a + 2b = 4$
$$11a + 5b = 3$$

Solution

1. Transform both equations by multiplication so that the b-terms are the same.

$5(3a + 2b) = 5(4) \rightarrow 15a + 10b = 20$
$2(11a + 5b) = 2(3) \rightarrow \underline{22a + 10b = 6}$

2. Subtract similar terms.

$-7a \qquad = 14$

3. Solve the resulting equation.

$a = -2$

4. Substitute for a in either original equation to find the value of b.

$3(-2) + 2b = 4$
$-6 + 2b = 4$
$2b = 10$

5. The check is left for you. The solution is $(-2, 5)$.

$b = 5$

Solve each system by using multiplication with the addition-or-subtraction method.

16. $3t - 8z = -2$
$7t + 4z = 18$

17. $6a + 7c = 8$
$2a + 5c = 8$

18. $4x + 9y = 3$
$-7x + 3y = -24$

19. $2x - 3y = 18$
$3x + 4y = -7$

20. $4x + 3y = -14$
$6x - 2y = -8$

21. $3a + 4b = 4$
$2a - 3b = 14$

22. $5m - 2n = -1$
$4m + 5n = -14$

23. $2x + 7y = 5$
$3x - 5y = 23$

24. $4x - 3y = 10$
$5x + 6y = -7$

25. $2x + 3y = 9$
$3x + 5y = 16$

26. $5x - 4y = 5$
$2x + 3y = 25$

27. $5a - 2c = 1$
$4a + 5c = 47$

28. $6x - 5y = 12$
$8x - 3y = 16$

29. $7x - 5y = 20$
$3x + 2y = 21$

30. $6x + 5y = 13$
$5x + 9y = 6$

31. $3x + 2y = 4$
$11x + 5y = 3$

32. $2x + 7y = -3$
$3x + 5y = 1$

33. $4x - 5y = 3$
$3x + 2y = -15$

Mixed Review Exercises

Factor completely.

1. $4 - 16x + 16x^2$

2. $6m^2n - 18mn^3$

3. $9c^2 - 16d^2$

4. $x^2 + 7x + 10$

5. $2y^2 + 7y + 3$

6. $p^2 - 5p - 14$

Find the constant of variation.

7. y varies directly as x, and $y = 63$ when $x = 9$.

8. t varies directly as s, and $t = -24$ when $s = 96$.

9. p is directly proportional to n, and $p = 27$ when $n = 36$.

10. h is directly proportional to k, and $h = 30$ when $k = 6$.

9–6 *Wind and Water Current Problems*

Objective: To use systems of equations to solve wind and water current problems.

Example   A jet can travel the 6000 km distance between Washington, D.C. and London in 6 h with the wind. The return trip against the same wind takes 7 h 30 min. Find the rate of the jet in still air and the rate of the wind.

Solution

Step 1   The problem asks for the rate of the jet in still air and the rate of the wind.

Step 2   Let r = the rate in km/h of the jet in still air.
Let w = the rate in km/h of the wind.

The time 7 h 30 min is $7\frac{30}{60}$ h, or 7.5 h.

	Rate	× Time	= Distance
With the wind	$r + w$	6	6000
Against the wind	$r - w$	7.5	6000

Step 3   Use the information in the chart to write two equations.

$$6(r + w) = 6000, \quad \text{or } r + w = 1000$$
$$7.5(r - w) = 6000, \quad \text{or } r - w = 800$$

Step 4
$$
\begin{aligned}
r + w &= 1000 \\
\underline{r - w} &= \underline{800} \\
2r &= 1800 \\
r &= 900
\end{aligned}
$$

$$
\begin{aligned}
900 + w &= 1000 \\
w &= 100
\end{aligned}
$$

Step 5   The check is left for you.

The rate of the jet is 900 km/h.
The rate of the wind is 100 km/h.

Solve.

1. A small plane traveled the 1200 km distance between two islands in 4 h with the wind. The return trip against the same wind took 5 h. Find the rate of the plane in still air and the rate of the wind.

2. A plane traveled the 2080 km distance between two cities in 5 h with the wind. The return trip against the same wind took 6.5 h. Find the rate of the plane in still air and the rate of the wind.

Study Guide, ALGEBRA, Structure and Method, Book 1

–6   *Wind and Water Current Problems*   (continued)

lve.

. A small plane can travel the 3200 km distance between two cities in 10 h with the wind. Against the same wind the plane can only fly 2400 km in 10 h. Find the rate of the plane in still air and the rate of the wind.

. A plane can fly 4800 km in 4 h with the wind. The return trip against the same wind takes 5 h. Find the rate of the plane in still air and the rate of the wind.

. The 4200 km trip from New York to San Francisco takes 7 h flying against the wind, but only 6 h returning. Find the speed of the plane in still air and the wind speed.

. Paddling with current, a canoeist can travel 48 km in 3 h. Against the current the canoeist takes 4 h to travel the same distance. Find the rate of the canoeist in still water and the rate of the current.

. A cabin cruiser traveling with the current went 120 km in 3 h. Against the current it took 5 h to travel the same distance. Find the rate of the cabin cruiser in still water and the rate of the current.

. A sailboat travels 24 mi downstream in 3 h. The return trip upstream takes 4 h. Find the speed of the sailboat in still water and the rate of the current.

. A crew can row 45 km downstream in 3 h. Rowing against the same current, the crew rowed the same distance in 5 h. Find the rowing rate of the crew in still water and the rate of the current.

. The 3600 km trip between two cities takes 6 h flying with the wind and 7.2 h against the wind. Find the speed of the plane in still air and the wind speed.

ixed Review Exercises

lve each system using multiplication with the dition-or-subtraction method.

. $2x - 3y = 12$
   $x + y = 1$

2. $x + y = 3$
   $3x - 5y = 17$

3. $2x - 3y = 6$
   $3x + 4y = -25$

mplify.

. $\dfrac{2n^2 - 13n + 20}{2n - 5}$

5. $\dfrac{3}{x - 1} + \dfrac{4}{1 - x}$

. $\dfrac{2x + 1}{6} - \dfrac{x + 3}{4}$

7. $a - 1 - \dfrac{a + 2}{a - 3}$

. $\dfrac{x^2 - 10xy + 25y^2}{x - y} \div \dfrac{x^2 - 4xy - 5y^2}{x^2 - y^2}$

NAME _____   DATE _____

9-7 Puzzle Problems

Objective: To use systems of equations to solve digit, age, and fraction problems.

Example 1 (Digit problem)

The sum of the digits in a two-digit number is 10. The new number obtained when the digits are reversed is 18 more than the original number. Find the original number.

Solution

Step 1 The problem asks for the original number.

Step 2 Let t = the tens digit and u = the units digit of the original number.

	Tens	Units	Value
Original number	t	u	$10t + u$
Number with digits reversed	u	t	$10u + t$

Step 3 Use the facts of the problem to write two equations.

$$t + u = 10 \quad \text{Sum of the digits of the original number is 10.}$$
$$(10u + t) - (10t + u) = 18 \quad \left\{ \begin{array}{l} \text{Difference between new number} \\ \text{and original number is 18.} \end{array} \right.$$
$$10u + t - 10t - u = 18$$
$$9u - 9t = 18$$
$$9(u - t) = 18$$
$$u - t = 2$$

Step 4
$$\begin{array}{l} u + t = 10 \\ \underline{u - t = 2} \\ 2u = 12 \\ u = 6 \end{array} \quad \left\{ \begin{array}{l} \text{Write the two equations as a system and} \\ \text{solve for one variable.} \end{array} \right.$$

$$u - t = 2 \quad \text{Substitute 6 for } u \text{ in the second equation.}$$
$$6 - t = 2$$
$$t = 4$$

Step 5 The check is left for you.

The original number is 46.

Solve by using a system of two equations in two variables.

1. The sum of the digits in a two-digit number is 7. The new number obtained when the digits are reversed is 27 less than the original number. Find the original number.

2. A two-digit number is seven times the sum of its digits. The tens digit is 3 more than the units digit. What is the number?

Study Guide, ALGEBRA, Structure and Method, Book 1
Copyright © by Houghton Mifflin Company. All rights reserved.

Example 2   **(Age problem)**   Chan is three years older than Myra. Six years ago Chan was twice as old as Myra was. Find their ages now.

Solution

Steps 1, 2   Let c = Chan's age now and let m = Myra's age now.

Step 3   Use the facts of the problem to write two equations:

$$c = m + 3 \qquad \text{\{now}$$

$$c - 6 = 2(m - 6) \qquad \text{\{six years ago}$$

Age	Now	6 years ago
Chan	c	$c - 6$
Myra	m	$m - 6$

Step 4   Simplify the equations and solve the system: $m = 9$, $c = 12$

Step 5   The check is left for you. Chan is 12 years old now and Myra is 9.

Example 3   **(Fraction problem)**   The denominator of a fraction is 4 more than the numerator. If 2 is subtracted from each, the value of the new fraction is $\frac{1}{5}$. Find the original fraction.

Solution

Steps 1, 2   Let n = the numerator and d = the denominator of the original fraction.

Step 3   Use the facts of the problem to write two equations.

$$d = n + 4$$
$$\frac{n - 2}{d - 2} = \frac{1}{5}, \text{ or } 5(n - 2) = d - 2$$

Step 4   Simplify the equations and solve the system: $n = 3$, $d = 7$

Step 5   The check is left to you. The original fraction $\frac{n}{d}$ is $\frac{3}{7}$.

Solve by using a system of two equations in two variables.

. Max is 5 years older than Paulette. Next year he will be twice as old as she will be. How old is each now?

. Gloria is 20 years older than Reggie. Five years ago she was five times as old as he was. How old is each now?

. The denominator of a fraction is 7 more than the numerator. If 5 is added to both the numerator and denominator, the value of the resulting fraction is $\frac{1}{2}$. What is the original fraction?

. The denominator of a fraction is 1 more than the numerator. If the numerator is decreased by 1, the value of the resulting fraction is $\frac{3}{4}$. What is the original fraction?

Mixed Review Exercises

Solve.

1. $\dfrac{x + 3}{2} - \dfrac{x}{3} = \dfrac{5}{6}$
   2. $\dfrac{3n + 2}{5} = \dfrac{n - 2}{3}$
   3. $\dfrac{3 + c}{2 + c} = \dfrac{3}{4}$
   4. $-\dfrac{1}{4}x = 12$

10 Inequalities

10–1 Order of Real Numbers

Objective: To review the concept of order and to graph inequalities in one variable.

Vocabulary

Inequality A statement formed by placing an inequality symbol between numerical or variable expressions.

Solutions of an inequality The values of the domain of the variable for which the inequality is true.

Solution set of an inequality The set of all solutions of the inequality.

Graph of an inequality The graph of the numbers in the solution set of an inequality.

Symbols

Inequality symbols Symbols used to show the order of two real numbers:

$x > 2$	(x is greater than 2.)	$x \geq 2$	(x is greater than *or* equal to 2.)
$x < 2$	(x is less than 2.)	$x \leq 2$	(x is less than *or* equal to 2.)

$-3 < x < 1$ (x is greater than -3 *and* less than 1.)

Example 1   Translate the statements into symbols.

   a. -2 is greater than -6 **b.** x is less than or equal to 5.

Solution   **a.** $-2 > -6$ **b.** $x \leq 5$

Translate the statements into symbols.

1. -2 is less than 5.
2. -3 is greater than -4.
3. -6 is less than or equal to -2.
4. 4 is greater than or equal to 1.
5. 4 is greater than 1 and less than 5.5.
6. 0 is greater than -2 and less than 3.
7. -5 is between 1 and -7.
8. 3 is between -5 and 5.
9. 3.5 is greater than 3 and 3 is greater than 0.
10. -3.5 is less than -3 and -3 is less than 0.
11. The number n is greater than 6.
12. The number n is less than 12.

Example 2   Classify each statement as true or false.

   a. $-2 < 2 < 4$ **b.** $-1 < 5 < 3$

Solution   **a.** $-2 < 2 < 4$ is true since *both* $-2 < 2$ and $2 < 4$ are true.

   b. $-1 < 5 < 3$ is false because $-1 < 5$ is true *but* $5 < 3$ is false.

–1 Order of Real Numbers *(continued)*

Example 3   Classify each statement as true or false.

a. $5 \le 8$ **b.** $2 \le 2$

Solution   **a.** For $5 \le 8$ to be true, *either* $5 < 8$ or $5 = 8$ must be true.

$5 \le 8$ is true since $5 < 8$.

b. $2 \le 2$ is true since $2 = 2$ is true.

---

classify each statement as true or false.

$-5 < 1 < 6$	**14.** $-8 < 2 < 5$	**15.** $3 > 0 > 1$	**16.** $-3 < -2 < 3$
$8 \ge 4$	**18.** $13 \le 22$	**19.** $-9 \le -16$	**20.** $-2 \ge -3$
$|-2| \ge -2$	**22.** $|-2| \le 0$	**23.** $|0.5| < -0.3$	**24.** $|-5| \le |-10|$

Example 4   Solve $y + 2 \le 3$ if $y \in \{-2, -1, 0, 1, 2, 3\}$.

Solution   Find all the values in the domain that make the inequality true.

Replace y with each of its values in turn:

$-2 + 2 \le 3$	**True**
$-1 + 2 \le 3$	**True**
$0 + 2 \le 3$	**True**
$1 + 2 \le 3$	**True**
$2 + 2 \le 3$	**False**
$3 + 2 \le 3$	**False**

The solution set is $\{-2, -1, 0, 1\}$

---

solve each inequality if $x \in \{-3, -2, -1, 0, 1, 2, 3\}$.

$2x < 4$	**26.** $3x < 3$	**27.** $-2x \le 6$	**28.** $x + 1 < 3$
$-2 + x \le 0$	**30.** $1 - x \ge 0$	**31.** $x^2 \ge 8$	**32.** $x^2 \le 9$

mixed Review Exercises

solve.

$x - 4 = 11$	**2.** $12 = 3(c - 1)$	**3.** $3 - 2a = 15$
$\dfrac{x}{2} = -15$	**5.** $\dfrac{24}{y} = \dfrac{8}{3}$	**6.** $\dfrac{2x + 4}{4} = \dfrac{x + 8}{3}$
$3(4 + n) = 2(n - 5)$	**8.** $(x + 2)(x + 5) = (x + 3)^2$	**9.** $\dfrac{n}{3} + 6 = n$
$\dfrac{x}{3} = \dfrac{x - 5}{4}$	**11.** $\dfrac{3x}{8} + \dfrac{x}{4} = \dfrac{5}{4}$	**12.** $\dfrac{1}{2}(x + 6) = 8$

10–2 Solving Inequalities

Objective: To transform inequalities in order to solve them.

Properties

Property of Comparison For all real numbers a and b, one and only one of the following statements is true: $a < b$, $a = b$, $a > b$.

Transitive Property of Order For all real numbers a, b, and c,
1. If $a < b$ and $b < c$, then $a < c$;
2. If $c > b$ and $b > a$, then $c > a$.

Addition Property of Order For all real numbers a, b, and c,
1. If $a < b$, then $a + c < b + c$;
2. If $a > b$, then $a + c > b + c$.

Multiplication Property of Order
For all real numbers a, b, and c, such that $c > 0$:
1. If $a < b$, then $ac < bc$;
2. If $a > b$, then $ac > bc$.
For all real numbers a, b, and c, such that $c < 0$:
1. If $a < b$, then $ac > bc$;
2. If $a > b$, then $ac < bc$.

Vocabulary

Equivalent inequality An inequality with the same solution set as another inequality.

Transformations That Produce an Equivalent Inequality

1. **Substituting** for either side of the inequality an expression equivalent to that side.

2. **Adding to (or subtracting from)** each side of the inequality the same real number.

3. **Multiplying (or dividing)** each side of the inequality by the same *positive* number.

4. **Multiplying (or dividing)** each side of the inequality by the same *negative* number and *reversing the direction of the inequality.*

CAUTION   Multiplying both sides of an inequality by zero does not produce an inequality; the result is the identity $0 = 0$.

Example 1	Tell how to transform the first inequality into the second one.

      a. $m - 6 > 2$ **b.** $-6k \geq 18$
            $m > 8$ $k \leq -3$

Solution   **a.** Add 6 to each side.   **b.** Divide each side by -6 and reverse the direction of the inequality.

Tell how to transform the first inequality into the second one.

1. $t + 2 < 6$ **2.** $x - 3 > 7$ **3.** $x + 5 < 0$
     $t < 4$ $x > 10$ $x < -5$

10–2 *Solving Inequalities* (continued)

Tell how to transform the first inequality into the second one.

4. $4p < 28$ **5.** $2m < -12$ **6.** $-7a < 21$ **7.** $3 < \dfrac{x}{5}$ **8.** $\dfrac{x}{-2} \le -4$ **9.** $-\dfrac{t}{3} \ge 0$
  $p < 7$ $m < -6$ $a > -3$ $15 < x$ $x \ge 8$ $t \le 0$

Example 2 Solve $4x - 1 < 7 + 2x$ and graph its solution set.

Solution
$$4x - 1 + 1 < 7 + 2x + 1 \qquad \text{Add 1 to each side.}$$
$$4x < 8 + 2x$$
$$4x - 2x < 8 + 2x - 2x \qquad \text{Subtract } 2x \text{ from each side.}$$
$$2x < 8$$
$$\dfrac{2x}{2} < \dfrac{8}{2} \qquad \text{Divide each side by 2.}$$
$$x < 4 \qquad \text{The solution set is \{the real numbers less than 4\}.}$$

The graph is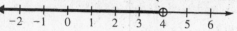

Example 3 Solve $2(w - 6) \ge 3(1 - w)$ and graph its solution set.

Solution
$$2w - 12 \ge 3 - 3w \qquad \text{Use the distributive property.}$$
$$5w \ge 15 \qquad \text{Add } 3w \text{ to each side and add 12 to each side.}$$
$$w \ge 3 \qquad \text{Divide each side by 5.}$$

The solution set is {the real numbers greater than or equal to 3}.

The graph is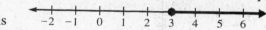

Solve each inequality. Graph the solution set.

10. $x - 2 \ge 3$ **11.** $8 < z + 2$ **12.** $4p < 20$ **13.** $15 \le 5w$

14. $-24 > -6m$ **15.** $\dfrac{d}{2} > -3$ **16.** $3 - g > 0$ **17.** $2v + 1 > 9$

18. $6 \ge 2k - 6$ **19.** $3 + \dfrac{x}{2} \le 4$ **20.** $6 - \dfrac{2}{3}c > 0$ **21.** $3r - 4 < 4r + 1$

22. $4y < 3y + 6$ **23.** $3f - 2 < 2f + 3$ **24.** $2r - 3 < 3r + 1$ **25.** $6 - 2b > 3 - b$

26. $2(x - 3) \le 4$ **27.** $6 < 3(2 - m)$ **28.** $3(x + 2) \le 3x + 2$ **29.** $4(k - 3) \ge 6(1 - k)$

Mixed Review Exercises

Classify each statement as true or false.

1. $|-2| > -(-1)$ **2.** $|-4| \le |4|$ **3.** $|-7| > |-8|$

Solve.

4. $5f - 3 = f + 17$ **5.** $0 = 3x + 12$ **6.** $3y - 2(y - 1) = -4$

7. $x - 2(8 - x) = -x$ **8.** $a(a + 4) = (a - 6)(a - 5)$ **9.** $3x + 2(x - 1) = x + 22$

10–3 Solving Problems Involving Inequalities

Objective: To solve problems involving inequalities.

Example 1 The sum of two consecutive integers is less than 80. Find the pair of such integers with the greatest sum.

Solution

Step 1   The problem asks for the largest pair of consecutive integers whose sum is less than 80.

Step 2   Let n = the smaller of the two consecutive integers.
Then $n + 1$ = the larger of the two consecutive integers.

Step 3   Use the variables to write an inequality based on the given information.
The sum of two consecutive integers is less than 80.
$$n + (n + 1) < 80$$

Step 4   Solve the open sentence:
$$n + n + 1 < 80$$
$$2n + 1 < 80$$
$$2n < 79$$
$$n < 39\tfrac{1}{2}$$

The largest integer less than $39\tfrac{1}{2}$ is 39. Thus, $n = 39$ and $n + 1 = 40$.

Step 5   *Check:* Is the sum $39 + 40$ less than 80?
$$39 + 40 \overset{?}{<} 80$$
$$79 < 80 \ \checkmark$$

39 and 40 form the largest pair of consecutive integers whose sum is less than 80.

For each of the following:
a. **Choose a variable to represent the number in bold face type.**
b. **Use the variable to write an inequality based on the given information.**
   (Do not solve.)

1. Harry, who is not yet 16 years old, is three years younger than Lena.
   (Lena's age)

2. After driving 125 miles, Barry still has more than 75 miles to travel.
   (the total number of miles Barry will drive)

Example 2 Translate each phrase into mathematical terms.

	Solution
a. The age of the house *is at least* 75 years	**a.** $a \geq 75$
b. The distance *is no less than* 250 km	**b.** $d \geq 250$
c. The price of the ticket *is at most* $190	**c.** $p \leq 190$
d. Her driving time to school *is no more than* 30 min	**d.** $t \leq 30$

0–3 Solving Problems Involving Inequalities (continued)

r each of the following:
Choose a variable to represent the number in bold face type.
Use the variable to write an inequality based on the given information.
(Do not solve.)

. Katrina's balance in her checking account is $160. She must deposit at least
enough money in her account to be able to pay her car payment of $295.
(the amount of deposit)

. Dan bicycled 12 more kilometers than one third **the number of kilometers
Manuel bicycled.** Dan bicycled at most 24 km.

. The length of a rectangle is 7 cm longer than **the width.** The perimeter is
no more than 38 cm.

. The sum of two consecutive odd integers is at most 185.
(the greater integer)

. The product of two consecutive integers is no less than 75.
(the smaller integer)

lve.

. The sum of two consecutive integers is less than 100. Find the pair of
integers with the greatest sum.

. The sum of two consecutive even integers is at most 180. Find the pair of
integers with the greatest sum.

. After selling 160 copies of the program to a school play, an usher had
fewer than 40 copies left. How many copies of the program did the usher
have originally?

. A house and a lot together cost more than $86,000. The house costs
$2000 more than six times the cost of the lot. How much does the lot cost
by itself?

. Andrew's salary is $1200 a month plus a 4% commission on all his sales.
What must the amount of his sales be to earn at least $1600 each month?

ixed Review Exercises

ve.

$|x| = 5$ 2. $|1 - 5| = k$ 3. $|y| - 2 = 6$

$2|b| = 16$ 5. $x = |-1 - (-3)|$ 6. $n = -|5 - 8|$

:tor completely.

$x^2 + 12x + 35$ 8. $x^3 - 3x^2 - 18x$ 9. $36x^2 - 25$

$2y^2 - 5y - 3$ 11. $x^2 + 8xy + 16y^2$ 12. $12x^3 - 3x$

10–4 Solving Combined Inequalities

Objective: To find the solution sets of combined inequalities.

Vocabulary

Conjunction A sentence formed by joining two open sentences by the word *and*. For example, $-1 < x$ *and* $x < 4$, which can also be written as $-1 < x < 4$.

Solve a conjunction To find the values of the variables for which *both* open sentences in the conjunction are true.

Disjunction A sentence formed by joining two open sentences by the word *or*. For example, $y > 1$ *or* $y = 1$.

Solve a disjunction To find the values of the variables for which *at least one* of the open sentences in the disjunction is true.

Example 1   Draw the graph of each open sentence.   **Solution**

   a. $4 < x$ **a.**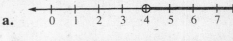

   b. $x < 6$ **b.**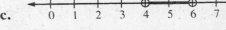

   c. conjunction: $4 < x$ *and* $x < 6$ **c.**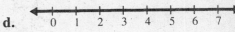

   d. disjunction: $4 < x$ *or* $x < 6$ **d.**

Draw the graph of each open sentence.

   1. $-2 < t$ and $t \le 1$ **2.** $r > 2$ or $r \le -1$ **3.** $2 \le n$ and $n \le 6$ **4.** $x < -1$ or $x \ge 1$

Example 2   Describe the graph of each open sentence.

    a. conjunction: $t < 3$ *and* $t \ge 3$ **b.** disjunction: $t < 3$ *or* $t \ge 3$.

Solution   **a.** No real number can be less than 3 and also greater than or equal to 3. The solution set is the empty set. It has no graph.

      b. Every real number is either less than 3 or greater than or equal to 3. The solution set is {the real numbers}. Its graph is the entire number line.

Example 3   Solve the conjunction $-2 \le x - 1 < 3$ and graph its solution set.

Solution 1   Solve the conjunction:

$$
\begin{array}{ccc}
-2 \le x - 1 & \text{and} & x - 1 < 3 \\
-2 + 1 \le x - 1 + 1 & | & x - 1 + 1 < 3 + 1 \\
-1 \le x & \text{and} & x < 4 \\
& -1 \le x < 4 &
\end{array}
$$

The solution set is $\{-1, \text{ and all the real numbers } between -1 \text{ and } 4\}$.

The graph is

0–4 Solving Combined Inequalities (continued)

Solution 2

$$-2 \leq \quad x - 1 \quad < 3$$
$$-2 + 1 \leq x - 1 + 1 < 3 + 1 \quad \text{Add 1 to each part of the inequality.}$$
$$-1 \leq \quad x \quad < 4$$

Example 4   Solve the disjunction $2x + 1 < 5$ or $3x \geq x + 8$ and graph its solution set.

Solution

$$
\begin{array}{ccc}
2x + 1 < 5 & \text{or} & 3x \geq x + 8 \\
2x + 1 - 1 < 5 - 1 & & 3x - x \geq x + 8 - x \\
2x < 4 & & 2x \geq 8 \\
x < 2 & \text{or} & x \geq 4
\end{array}
$$

The solution set is {4, and the real numbers greater than 4 *or* less than 2}.

The graph is

Solve each open sentence. Graph the solution set, if there is one.

5. $-1 < a - 1 < 4$

6. $-3 < y + 1 \leq 2$

7. $-2 < -3 + d \leq 1$

8. $-4 \leq 2 + r < 2$

9. $-4 \leq 2a + 6 < 10$

10. $-3 < 2b + 1 \leq 5$

11. $-8 \leq 3m + 1 < 7$

12. $-4 < 3n + 5 \leq 8$

13. $x - 1 < -4$ or $x - 1 > 5$

14. $h + 3 \leq -1$ or $h + 3 \geq 1$

15. $2x - 1 \leq -5$ or $2x - 1 > 5$

16. $3 + 2y < -5$ or $3 + 2y > 5$

17. $-5x > 20$ or $10 + 5x \geq 0$

18. $2d - 3 < -5$ or $5 < 2d - 3$

19. $-3m < 6$ and $18 + 3m < 0$

20. $-3 \leq 1 - t$ and $1 - t < 2$

Mixed Review Exercises

Choose a variable and use the variable to write an inequality.

1. The finish line is at least 20 yd away.

2. The temperature cannot exceed 25 °C.

3. The weight is at most 105 lb.

4. The flight takes at least 2 h.

5. The cost is not more than $75.

6. The tolerance is smaller than 1 cm.

7. Ray averages at most 15 points per game.

8. Joy won at least 12 tennis matches.

Evaluate each expression if $k = -3$, $m = 9$, and $x = 3$.

9. $|x - k|$

10. $|m - k|$

11. $|x + k|$

12. $|k - x|$

13. $|k - m|$

14. $|k + m|$

10–5 *Absolute Value in Open Sentences*

Objective: To solve equations and inequalities involving absolute value.

Symbols

$|a - b| = |b - a|$ (The distance between a and b on a number line.)

$|a + b| = |a - (-b)|$ (The distance between a and the opposite of b on a number line.)

Example 1   Solve $|x - 1| = 2$.

Solution 1   To satisfy $|x - 1| = 2$, x must be a number whose distance from 1 is 2.
To arrive at x, start at 1 and move 2 units in either direction on the number line.

You arrive at 3 and -1 as the values of x. The solution set is $\{-1, 3\}$.

Solution 2   Note that $|x - 1| = 2$ is equivalent to the disjunction:

$$x - 1 = -2 \quad \text{or} \quad x - 1 = 2$$
$$x = -1 \quad \text{or} \quad x = 3 \qquad \text{The solution set is } \{-1, 3\}.$$

Solve.

1. $|m - 3| = 5$ 2. $|k + 4| = 1$ 3. $|2 + x| = 4$

4. $|7 - x| = 3$ 5. $|x - 5| = 2$ 6. $|6 - x| = 7$

Example 2   Solve $|x + 2| \leq 4$ and graph its solution set.

Solution 1 $|x + 2| \leq 4$ is equivalent to $|x - (-2)| \leq 4$.
The distance between x and -2 must be no more than 4.

Starting at -2, numbers within 4 units in either direction will satisfy $|x + 2| \leq 4$.
Thus, $|x + 2| \leq 4$ is equivalent to $-6 \leq x \leq 2$.

The solution set is $\{-6, 2, \text{and the real numbers between } -6 \text{ and } 2\}$.
The graph is shown above.

Solution 2 $|x + 2| \leq 4$ is equivalent to the conjunction:

$$-4 \leq \quad x + 2 \quad \leq 4$$
$$-4 - 2 \leq x + 2 - 2 \leq 4 - 2 \qquad \left\{ \begin{array}{l} \text{The solution set and graph} \\ \text{are as in Solution 1.} \end{array} \right.$$
$$-6 \leq \quad x \quad \leq 2$$

10–5 Absolute Value in Open Sentences (continued)

Example 3 Solve $|t - 2| > 3$ and graph its solution set.

Solution 1 The distance between t and 2 must be greater than 3, as shown below:

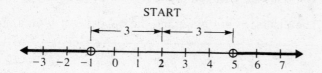

Therefore, $|t - 2| > 3$ is equivalent to the disjunction

$$t < -1 \quad \text{or} \quad t > 5.$$

The solution set is {the real numbers less than -1 or greater than 5}. The graph is shown above.

Solution 2 $|t - 2| > 3$ is equivalent to the disjunction:

$$t - 2 < -3 \quad \text{or} \quad t - 2 > 3$$
$$t < -1 \quad \text{or} \quad t > 5$$

The solution set and graph are as in Solution 1.

Solve each open sentence and graph its solution set.

7. $|x| > 2$ 8. $|x| \leq 2$ 9. $|x| \geq 1$

10. $|x - 2| < 1$ 11. $|x - 2| > 2$ 12. $|x + 2| \geq 1$

13. $|x - 1| \leq 1$ 14. $|x - 1| \geq 1$ 15. $|x + 3| \leq 1$

16. $|x + 1| > 1$ 17. $|x - 3| \geq 4$ 18. $|x - 4| < 2$

19. $|3 - v| \geq 5$ 20. $|2 - x| \geq 1$ 21. $|-2 - x| \leq 4$

Mixed Review Exercises

Solve each inequality and graph its solution set.

1. $x - 3 < 5$ 2. $\dfrac{x}{3} + 6 < 2$ 3. $8 < 4(3 + m)$

4. $-1 < x + 4 < 1$ 5. $h + 2 \leq 8 \text{ or } h - 3 > 2$ 6. $2 \leq -x \leq 8$

Simplify.

7. $\dfrac{15x}{4y^2} \div 3xy$ 8. $\left(\dfrac{4a}{b}\right) \cdot \left(\dfrac{5b}{2a}\right)^2$

9. $\dfrac{x + 2}{3} - \dfrac{2x}{6}$ 10. $2x + \dfrac{x}{5}$

10–6 Absolute Values of Products in Open Sentences

Objective: To extend your skill in solving open sentences that involve absolute value.

Property

The absolute value of a product of numbers equals the product of their absolute values.

$$|ab| = |a| \cdot |b|$$

Examples: $|-3 \cdot 5| = |-15| = 15 = 3 \cdot 5 = |-3| \cdot |5|$

$|-6 \cdot (-2)| = |12| = 12 = 6 \cdot 2 = |-6| \cdot |-2|$

Example 1   Solve $|2x + 1| = 5$.

Solution 1 $|2x + 1| = 5$ is equivalent to the disjunction:

$$
\begin{array}{lcl}
2x + 1 = -5 & \text{or} & 2x + 1 = 5 \\
2x + 1 - 1 = -5 - 1 & & 2x + 1 - 1 = 5 - 1 \\
2x = -6 & & 2x = 4 \\
x = -3 & \text{or} & x = 2
\end{array}
$$

The solution set is $\{-3, 2\}$.

Solution 2 $|2x + 1| = 5$

$$\left| 2\left(x + \tfrac{1}{2} \right) \right| = 5$$

$$|2| \cdot \left| x + \tfrac{1}{2} \right| = 5$$

$$2\left| x + \tfrac{1}{2} \right| = 5$$

$$\left| x + \tfrac{1}{2} \right| = \tfrac{5}{2}$$

$$\left| x - \left(-\tfrac{1}{2} \right) \right| = \tfrac{5}{2} \qquad \text{Thus the distance between } x \text{ and } -\tfrac{1}{2} \text{ is } \tfrac{5}{2}.$$

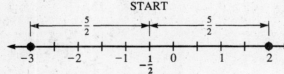

$\begin{cases} \text{Starting at } -\tfrac{1}{2} \text{ the numbers} \\ -3 \text{ and } 2 \text{ are exactly } \tfrac{5}{2} \text{ units} \\ \text{away in either direction.} \end{cases}$

The solution set is $\{-3, 2\}$.

Solve each open sentence and graph its solution set.

1. $|2y| = 6$

2. $|6y| = 24$

3. $|5x| = 10$

4. $\left| \dfrac{x}{3} \right| = 2$

5. $\left| \dfrac{x}{2} \right| = 4$

6. $|2a - 1| = 5$

7. $|2x + 1| = 7$

8. $|3x - 1| = 5$

9. $\left| \dfrac{x}{2} - 1 \right| = 3$

0–6   Absolute Values of Products in Open Sentences (continued)

Example 2   Solve $|8 - 2k| \geq 8$ and graph its solution set.

Solution 1
$$|8 - 2k| \geq 8$$
$$|-2k + 8| \geq 8$$
$$|(-2)(k - 4)| \geq 8 \qquad \text{Factor.}$$
$$|-2| \cdot |k - 4| \geq 8 \qquad \text{Use the property about the absolute value of a product.}$$
$$|k - 4| \geq 4$$

The distance between k and 4 must be 4 or more, as shown above.

Thus the given inequality is equivalent to the disjunction
$$k \leq 0 \qquad \text{or} \qquad k \geq 8$$

The solution set is $\{0, 8, \text{and the real numbers less than 0 or greater than 8}\}$.

The graph is shown above.

Solution 2 $|8 - 2k| \geq 8$ is equivalent to the disjunction
$$\begin{array}{lll} 8 - 2k \leq -8 & \text{or} & 8 - 2k \geq 8 \\ -2k \leq -16 & & -2k \geq 0 \\ k \geq 8 & \text{or} & k \leq 0 \end{array}$$

The solution set and graph are as given in Solution 1.

olve each open sentence and graph its solution set.

. $|2y - 1| \leq 5$ **11.** $|2x + 1| \geq 1$ **12.** $|2x - 3| < 7$

. $|2n - 1| \geq 3$ **14.** $|4x - 13| > 7$ **15.** $|6 - 3k| \geq 9$

. $|4 - 2k| \leq 4$ **17.** $\left|\dfrac{x}{2} - 1\right| \geq 3$ **18.** $\left|\dfrac{x}{3} - 2\right| \leq 2$

ixed Review Exercises

ve the slope and y-intercept of each line.

. $y = 3x + 1$ **2.** $3y = 12x - 6$ **3.** $3y - 2x + 6 = 0$

. $y = 6$ **5.** $2x - y = 5$ **6.** $x = -2y + 4$

aph each equation.

. $y = -x + 2$ **8.** $y = 2x - 3$ **9.** $x = -2$

. $y = 3$ **11.** $y = \dfrac{2}{3}x + 1$ **12.** $y = -\dfrac{1}{2}x - 2$

10–7 *Graphing Linear Inequalities*

Objective: To graph linear inequalities in two variables.

Vocabulary

Boundary A line that separates the coordinate plane into three sets of points: the points *on* the line, the points *above* the line, the points *below* the line.

If the boundary line is part of a graph, it is drawn as a *solid* line.
If the boundary line is *not* part of the graph, it is drawn as a *dashed* line.

Open half-plane Either of the two regions into which a boundary line separates the coordinate plane.

Closed half-plane The graph of an open half-plane and its boundary.

To graph a linear inequality in the variables x and y, when the coefficient of y is not zero:

1. **Transform** the given inequality into an equivalent inequality that has y alone as one side.

2. **Graph** the equation of the boundary. Use a solid line if the symbol $\geq$ or $\leq$ is used; use a dashed line if $>$ or $<$ is used.

3. **Shade** the appropriate region.

Example 1   Graph $2x - y \geq -4$.

Solution   1. Transform the inequality.

$$2x - y \geq -4$$
$$-y \geq -4 - 2x$$
$$y \leq 4 + 2x$$
$$y \leq 2x + 4$$

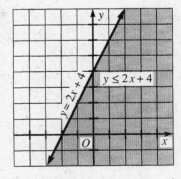

2. Draw the boundary line $y = 2x + 4$ as a *solid* line, since the symbol $\leq$ includes the equals sign.

3. Shade the region *below* the line since the symbol $\leq$ indicates the less than sign.

Check: Choose a point on the graph not on the boundary, say $(0, 0)$. See whether it satisfies the given inequality:

$$2x - y \geq -4$$
$$2(0) - 0 \geq -4$$
$$0 \geq -4 \ \checkmark$$

Thus, $(0, 0)$ is in the solution set, and the correct region has been shaded.

Example 2   Graph $y > 3$.

Solution   Graph $y = 3$ as a dashed horizontal line.

Any point above that line has a y-coordinate
that satisfies $y > 3$.

Therefore, the graph of $y > 3$ is the open
half-plane *above* the graph of $y = 3$.

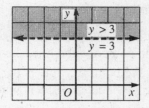

Example 3   Graph $x < -2$.

Solution   Graph $x = -2$ as a dashed vertical line.

Any point to the left of that vertical line has
an x-coordinate that satisfies $x < -2$.

Therefore, the graph of $x < -2$ is the open
half-plane *to the left* of the graph of $x = -2$.

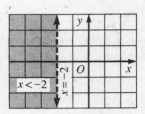

Graph each inequality.

1. $y \geq 2$ **2.** $y > 2$ **3.** $x < 2$ **4.** $x \leq 2$

5. $x > 1$ **6.** $y < 3$ **7.** $y \leq -1$ **8.** $y \leq 3$

9. $y < x + 3$ **10.** $y > -x + 2$ **11.** $y \leq 4 - x$ **12.** $y \geq 1 - 2x$

**Transform each inequality into an equivalent inequality with y as one
side. Then graph the inequality.**

13. $x + y > 1$ **14.** $x - y \geq 2$ **15.** $x - 2y \leq -4$

16. $2x + y > -2$ **17.** $3x - y > 6$ **18.** $y - 2x \leq -6$

19. $2x - 3y \geq 6$ **20.** $2y - 3x < 0$ **21.** $3y - 5 > 2(x + 2y)$

22. $2y - 1 > 3x - 5$ **23.** $3(x - y) \geq 2x + 1$ **24.** $4y - 6 < 2(x + y)$

Mixed Review Exercises

Solve each system by whatever method you prefer.

1. $y = 2x$ **2.** $m + n = 7$ **3.** $8p + q = -6$
   $x - y = 1$ $m - n = 3$ $8p - 6q = -20$

Solve each open sentence and graph its solution set.

4. $|3p| = 12$ **5.** $|2p + 2| = 10$ **6.** $|2x| < 12$

7. $|2x + 3| \geq 7$ **8.** $-5 \leq x + 1 < 4$ **9.** $2x + 3 > 5$ or $3 - x \geq 1$

10–8 *Systems of Linear Inequalities*

Objective: To graph the solution set of a system of two linear inequalities in two variables.

Example   Graph the solution set of the system:

$$y - x - 2 \leq 0$$
$$3x + 2y > -6$$

Solution   1. **Transform** each inequality into an equivalent one with y as one side.

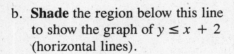

$$y - x - 2 \leq 0 \quad \longrightarrow \quad y \leq x + 2$$
$$3x + 2y > -6 \quad \longrightarrow \quad y > -\frac{3}{2}x - 3$$

2. a. **Draw** the graph of $y = x + 2$, the boundary for $y \leq x + 2$.

Use a solid line because the inequality has a $\leq$.

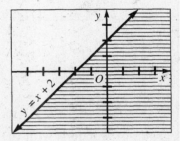

b. **Shade** the region below this line to show the graph of $y \leq x + 2$ (horizontal lines).

3. a. In the same coordinate system, **draw** the graph of $y = -\frac{3}{2}x - 3$, the boundary for $y > -\frac{3}{2}x - 3$.

Use a dashed line because the inequality has a $>$.

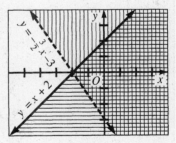

b. **Shade** the region above this line to show the graph of $y > -\frac{3}{2}x - 3$ (vertical lines).

4. The doubly shaded region (the intersection of the vertical and horizontal lines) is the graph of the solution set of the given system.

0–8   Systems of Linear Inequalities   *(continued)*

raph each pair of inequalities and indicate the solution set of the
stem with crosshatching.

1. $y > 0$
   $x \geq 0$

2. $y \leq 3$
   $x \geq 2$

3. $y > 2$
   $x < -1$

4. $y < -2$
   $x > 2$

5. $x < y$
   $y > 1$

6. $y > 2x$
   $x < 2$

7. $x \leq 2$
   $y > 3 - x$

8. $x > -1$
   $y \leq 2x + 5$

9. $y \leq x + 2$
   $y > 1 - x$

10. $y < 2x + 2$
    $y > -2x + 2$

11. $y > 2x - 1$
    $y < 2x + 2$

12. $y < 3x + 4$
    $y > 3 - 3x$

13. $x - y \geq 2$
    $x + y \leq 4$

14. $x + y \geq 3$
    $x - 2y > 4$

15. $2x - y > -1$
    $x - y > -2$

16. $x - y < 5$
    $x - 2y > 6$

17. $2x - 3y < -6$
    $2x + 3y < 0$

18. $2x - y > 0$
    $x - 2y \leq -6$

Iixed Review Exercises

ewrite each group of fractions with their LCD.

1. $\dfrac{1}{3}, \dfrac{8}{15}, \dfrac{2}{5}$

2. $\dfrac{a}{2}, \dfrac{3}{8}, \dfrac{a + 1}{12}$

3. $\dfrac{k}{k + 3}, \dfrac{2k}{k^2 + 6k + 9}$

4. $\dfrac{n + 2}{n - 4}, \dfrac{2}{n}, \dfrac{n}{4}$

5. $\dfrac{6}{x + 1}, \dfrac{x}{x - 1}$

6. $\dfrac{1}{x^2 - 4}, \dfrac{3}{2 - x}, \dfrac{x}{2 + x}$

Evaluate each expression if $a = \dfrac{2}{5}$, $b = \dfrac{1}{2}$, and $c = \dfrac{3}{10}$.

7. $a + b + c$

8. $c(b - a)$

9. $a - (b + c)$

10. $\dfrac{1}{3}(a + b + c)$

11. $a + \dfrac{1}{2}(b - c)$

12. $a - \dfrac{1}{2}(b - c)$

11 Rational Numbers

11–1 Properties of Rational Numbers

Objective: To learn and apply some properties of rational numbers.

Vocabulary

Rational number A real number that can be expressed as the quotient of two integers.

Examples: $\dfrac{2}{3}$, $6 = \dfrac{6}{1}$, $0 = \dfrac{0}{1}$, $4\dfrac{2}{3} = \dfrac{14}{3}$, $0.57 = \dfrac{57}{100}$, $-\dfrac{3}{5} = \dfrac{-3}{5}$

Average of two numbers The number halfway between two numbers.

Properties

Density Property of Rational Numbers Between every pair of different rational numbers there is another rational number.

Order Property of Rational Numbers For all integers a and b and all positive integers c and d:

1. $\dfrac{a}{c} > \dfrac{b}{d}$ if and only if $ad > bc$.   2. $\dfrac{a}{c} < \dfrac{b}{d}$ if and only if $ad < bc$.

Example 1   Which rational number is greater, $\dfrac{7}{3}$ or $\dfrac{9}{4}$?

Solution 1   The LCD is 12.

$\dfrac{7}{3} = \dfrac{28}{12}$ and $\dfrac{9}{4} = \dfrac{27}{12}$

Compare $\dfrac{28}{12}$ and $\dfrac{27}{12}$.

Since $28 > 27$, $\dfrac{28}{12} > \dfrac{27}{12}$. So $\dfrac{7}{3} > \dfrac{9}{4}$.

Solution 2 $\dfrac{7}{3} \overset{?}{-} \dfrac{9}{4}$

$(7)(4) \overset{?}{-} (3)(9)$

$28 > 27$

So $\dfrac{7}{3} > \dfrac{9}{4}$.

Arrange each group of numbers in order from least to greatest.

1. $\dfrac{3}{4}, \dfrac{5}{8}, \dfrac{4}{5}$

2. $\dfrac{2}{3}, \dfrac{11}{15}, \dfrac{5}{7}$

3. $4.6, \dfrac{105}{22}, -4$

4. $-\dfrac{31}{8}, -3.9, -\dfrac{41}{10}$

5. $-\dfrac{4}{9}, -\dfrac{6}{11}, -\dfrac{5}{7}, -\dfrac{3}{5}$

6. $\dfrac{5}{24}, \dfrac{4}{15}, \dfrac{5}{12}, \dfrac{1}{4}$

Example 2   Replace the $\underline{\ ?\ }$ with $<$, $=$, or $>$ to make a true statement.

a. $-\dfrac{3}{8} \overset{?}{-} -\dfrac{4}{11}$

b. $5\dfrac{3}{7} \overset{?}{-} \dfrac{49}{9}$

Solution   a. $\dfrac{-3}{8} \overset{?}{-} \dfrac{-4}{11}$

$(-3)(11) \overset{?}{-} (8)(-4)$ $\Big\{$ Use the order property

$-33 < -32$ of rational numbers.

So $-\dfrac{3}{8} < -\dfrac{4}{11}$.

b. $\dfrac{38}{7} \overset{?}{-} \dfrac{49}{9}$

$(38)(9) \overset{?}{-} (7)(49)$

$342 < 343$

So $5\dfrac{3}{7} < \dfrac{49}{9}$.

11-1 Properties of Rational Numbers (continued)

Replace the _?_ with < , = , or > to make a true statement.

7. $\frac{5}{6}$ _?_ $\frac{13}{18}$ 8. $\frac{3}{4}$ _?_ $\frac{11}{16}$ 9. $\frac{7}{8}$ _?_ $\frac{28}{32}$ 10. $\frac{2}{9}$ _?_ $\frac{13}{54}$

11. $\frac{1}{6}$ _?_ $\frac{5}{32}$ 12. $\frac{3}{4}$ _?_ $\frac{11}{15}$ 13. $-\frac{3}{5}$ _?_ $-\frac{7}{12}$ 14. $-\frac{7}{8}$ _?_ $-\frac{5}{6}$

15. $\frac{2}{5}$ _?_ $\frac{21}{50}$ 16. $-\frac{3}{4}$ _?_ $-\frac{8}{10}$ 17. $-12\frac{1}{5}$ _?_ $-\frac{86}{7}$ 18. $\frac{-107}{7}$ _?_ $-15\frac{5}{8}$

Example 3   Find the number halfway between $\frac{3}{8}$ and $\frac{2}{5}$.

Solution $\frac{3}{8} + \frac{1}{2}\left(\frac{2}{5} - \frac{3}{8}\right) = \frac{3}{8} + \frac{1}{2}\left(\frac{16}{40} - \frac{15}{40}\right)$ *Check:*   Is $\frac{3}{8} < \frac{31}{80}$?

$= \frac{3}{8} + \frac{1}{2}\left(\frac{1}{40}\right)$ Is $(3)(80) < (8)(31)$?

                           $240 < 248 \checkmark$

$= \frac{3}{8} + \frac{1}{80}$

$= \frac{30}{80} + \frac{1}{80}$ Is $\frac{31}{80} < \frac{2}{5}$?

$= \frac{31}{80}$ Is $(31)(5) < (80)(2)$?

                           $155 < 160 \checkmark$

$\frac{31}{80}$ is a rational number halfway between $\frac{3}{8}$ and $\frac{2}{5}$.

Find the number halfway between the given numbers.

19. $\frac{3}{8}, \frac{5}{9}$ 20. $\frac{7}{12}, \frac{5}{6}$ 21. $\frac{6}{11}, \frac{2}{3}$

22. $-\frac{11}{15}, -\frac{7}{12}$ 23. $-1\frac{2}{5}, -2\frac{5}{6}$ 24. $-2\frac{3}{5}, 4\frac{1}{3}$

Example 4   If $x \in \{0, 1, 2, 3\}$, state whether $\frac{x}{2}$ increases or decreases in value as x takes on its values in increasing order.

Solution   Replace x with 0, 1, 2, and 3 in order. $\frac{x}{2}$ becomes $\frac{0}{2}, \frac{1}{2}, \frac{2}{2}, \frac{3}{2}$. So $\frac{x}{2}$ increases.

If $x \in \{0, 1, 2, 3\}$, state whether each fraction increases or decreases in value as x takes on its value in increasing order.

25. $\frac{x}{6}$ 26. $-\frac{x}{8}$ 27. $\frac{4}{x+1}$ 28. $\frac{x+1}{4}$ 29. $\frac{7-2x}{5}$ 30. $\frac{12}{10-3x}$

Mixed Review Exercises

Solve each inequality and graph its solution.

1. $3y + 1 \le 7$ 2. $|0.2 + x| < 8$ 3. $6 + 4|2 - k| \ge 14$

4. $|x + 1| \ge 3$ 5. $6 \le 3x + 6 < 9$ 6. $5 \le 3 - 2m$

11–2 Decimal Forms of Rational Numbers

Objective: To express rational numbers as decimals or fractions.

Vocabulary

Terminating decimal The result when a common fraction is written as a decimal by dividing the numerator by the denominator and the remainder is zero. Also called *ending decimal* or *finite decimal*. For example, $\frac{3}{8} = 0.375$.

Nonterminating decimal The result when a common fraction is written as a decimal by dividing the numerator by the denominator and a digit or a block of digits repeat endlessly as the remainder. Also called *unending, infinite, repeating,* or *periodic decimals*. For example, $\frac{7}{11} = 0.6363 \ldots = 0.\overline{63}$. Dots or an overbar are used to indicate the repeating block of digits.

Example 1   Express $\frac{5}{8}$ as a decimal.

Solution

The division at the right shows that $\frac{5}{8}$ can be expressed as the terminating decimal 0.625.

$$
\begin{array}{r}
0.625 \\
8\overline{)5.000} \\
\underline{4\,8} \\
20 \\
\underline{16} \\
40 \\
\underline{40} \\
0
\end{array}
$$

Example 2   Express each rational number as a decimal: **a.** $\frac{1}{6}$ **b.** $\frac{2}{11}$ **c.** $2\frac{1}{7}$

Solution   If you don't reach a remainder of zero when dividing the numerator by the denominator, continue to divide until the remainders begin to repeat.

a. $\frac{1}{6} \longrightarrow$
$$
\begin{array}{r}
0.166 \\
6\overline{)1.000} \\
\underline{6} \\
40 \\
\underline{36} \\
40 \\
\underline{36} \\
4
\end{array}
$$

$\frac{1}{6} = 0.166\ldots = 0.1\overline{6}$

b. $\frac{2}{11} \longrightarrow$
$$
\begin{array}{r}
0.1818 \\
11\overline{)2.0000} \\
\underline{1\,1} \\
90 \\
\underline{88} \\
20 \\
\underline{11} \\
90 \\
\underline{88} \\
2
\end{array}
$$

$\frac{2}{11} = 0.1818\ldots = 0.\overline{18}$

c. $2\frac{1}{7} = \frac{15}{7} \longrightarrow$
$$
\begin{array}{r}
2.142857 \\
7\overline{)15.000000} \\
\underline{14} \\
1\,0 \\
\underline{7} \\
30 \\
\underline{28} \\
20 \\
\underline{14} \\
60 \\
\underline{56} \\
40 \\
\underline{35} \\
50 \\
\underline{49} \\
1
\end{array}
$$

$2\frac{1}{7} = 2.142857\ldots = 2.\overline{142857}$

1–2 Decimal Forms of Rational Numbers (continued)

Express each rational number as a terminating or repeating decimal.

1. a. $\frac{1}{3}$ b. $\frac{1}{30}$ 2. a. $\frac{5}{2}$ b. $\frac{5}{200}$ 3. a. $-\frac{2}{9}$ b. $-\frac{2}{9000}$ 4. a. $-\frac{2}{5}$ b. $-\frac{2}{50}$

5. $\frac{13}{8}$ 6. $\frac{5}{12}$ 7. $\frac{7}{27}$ 8. $-\frac{5}{18}$ 9. $3\frac{3}{20}$ 10. $2\frac{4}{11}$ 11. $-5\frac{3}{4}$ 12. $\frac{11}{27}$

Example 3 Express each terminating decimal as a fraction in simplest form.

a. 0.24   b. 0.325

Solution a. $0.24 = \frac{24}{100} = \frac{6}{25}$ b. $0.325 = \frac{325}{1000} = \frac{13}{40}$

Example 4 Express $0.5\overline{21}$ as a fraction in simplest form.

Solution Let $N =$ the number $0.5\overline{21}$ and $n =$ the number of digits in the block of repeating digits.

Multiply N by 10^n. Since $0.5\overline{21}$ has 2 digits in the repeating block, $n = 2$. Therefore, multiply both sides of the equation $N = 0.5\overline{21}$ by 10^2 or 100.

$$100N = 100(0.5\overline{21}).$$

Since $0.5\overline{21} = 0.52121\ldots$, $0.5\overline{21}$ can also be written as $0.521\overline{21}$.

Then $100(0.5\overline{21}) = 100(0.521\overline{21})$
                                    $= 52.1\overline{21}$

Subtract N from $100N$. $100N = 52.1\overline{21}$
                                      $N = 0.5\overline{21}$
                                      _____

Solve for N. $99N = 51.6$

$$N = \frac{51.6}{99} = \frac{516}{990} = \frac{86}{165} \quad \text{So } 0.5\overline{21} = \frac{86}{165}.$$

Express each rational number as a fraction in simplest terms.

13. 0.3   14. 0.88   15. 0.225   16. 2.6   17. 4.26

18. $0.\overline{2}$ 19. $1.8\overline{3}$ 20. $0.\overline{074}$ 21. $2.\overline{21}$ 22. $0.\overline{09}$

23. $0.\overline{37}$ 24. $0.\overline{63}$ 25. $0.08\overline{3}$ 26. $0.0\overline{8}$ 27. $-2.\overline{18}$

Mixed Review Exercises

Find the prime factorization of each number.

1. 120   2. 50   3. 484   4. 1125   5. 196   6. 288

Solve.

7. $(y + 2)(y - 3) = 0$ 8. $(a + 2)^2 = 16$ 9. $x^2 = -9$

10. $k^3 - 25k = 0$ 11. $|x + 2| = 6$ 12. $k + 3 < 12$

11–3 Rational Square Roots

Objective: To find the square roots of numbers that have rational square roots.

Vocabulary

Square root If $a^2 = b$, then a is a square root of b. Positive numbers have two square roots that are opposites. Example: Since $5^2 = 25$, 5 is a square root of 25. Since $(-5)^2 = 25$, -5 is also a square root of 25.

Radicand The symbol written beneath a radical sign.

Principal square root The positive square root of a positive number.

Symbols

$\sqrt{}$ (radical sign)

$\sqrt{9}$ (the positive square root of 9)

$-\sqrt{9}$ (the negative square root of 9)

$\pm\sqrt{9}$ (the positive or negative square root of 9)

Properties	Examples
Product Property of Square Roots For any nonnegative real numbers a and b: $$\sqrt{ab} = \sqrt{a} \cdot \sqrt{b}.$$	$\sqrt{4 \cdot 9} = \sqrt{4} \cdot \sqrt{9}$
Quotient Property of Square Roots For any nonnegative real number a and any positive real number b: $$\sqrt{\frac{a}{b}} = \frac{\sqrt{a}}{\sqrt{b}}.$$	$\sqrt{\frac{36}{4}} = \frac{\sqrt{36}}{\sqrt{4}}$

CAUTION Negative numbers do not have square roots in the set of real numbers. The square root of zero is zero.

Example 1 Find $\sqrt{256}$.

Solution
$$\begin{aligned}
\sqrt{256} &= \sqrt{4 \cdot 64} \\
&= \sqrt{4} \cdot \sqrt{64} \\
&= 2 \cdot 8 \\
&= 16
\end{aligned}$$

Example 2 Find $\sqrt{1764}$.

Solution
$$\begin{aligned}
\sqrt{1764} &= \sqrt{2^2 \cdot 3^2 \cdot 7^2} \\
&= \sqrt{2^2} \cdot \sqrt{3^2} \cdot \sqrt{7^2} \\
&= 2 \cdot 3 \cdot 7 \\
&= 42
\end{aligned}$$

{ If you cannot see any squares that divide into the radicand, begin by factoring the radicand.

NAME _____ DATE _____

1–3 *Rational Square Roots* (continued)

Find the indicated square root.

1. $\sqrt{49}$ **2.** $\sqrt{81}$ **3.** $\sqrt{144}$ **4.** $\sqrt{196}$

5. $-\sqrt{225}$ **6.** $-\sqrt{121}$ **7.** $\sqrt{576}$ **8.** $\sqrt{400}$

9. $\pm\sqrt{1600}$ **10.** $\pm\sqrt{2025}$ **11.** $\sqrt{900}$ **12.** $\sqrt{784}$

13. $\pm\sqrt{676}$ **14.** $\pm\sqrt{529}$ **15.** $-\sqrt{441}$ **16.** $\sqrt{484}$

Example 3 Find the indicated square root: **a.** $\sqrt{\dfrac{25}{81}}$ **b.** $\pm\sqrt{\dfrac{121}{289}}$

Solution **a.** $\sqrt{\dfrac{25}{81}} = \dfrac{\sqrt{25}}{\sqrt{81}} = \dfrac{5}{9}$ **b.** $\pm\sqrt{\dfrac{121}{289}} = \pm\dfrac{\sqrt{121}}{\sqrt{289}} = \pm\dfrac{11}{17}$

Find the indicated square root.

17. $\sqrt{\dfrac{81}{400}}$ **18.** $-\sqrt{\dfrac{225}{64}}$ **19.** $\pm\sqrt{\dfrac{121}{36}}$ **20.** $\sqrt{\dfrac{144}{625}}$

21. $\pm\sqrt{\dfrac{484}{529}}$ **22.** $-\sqrt{\dfrac{324}{361}}$ **23.** $\sqrt{\dfrac{225}{484}}$ **24.** $-\sqrt{\dfrac{289}{400}}$

25. $\pm\sqrt{\dfrac{64}{2025}}$ **26.** $\sqrt{\dfrac{256}{1225}}$ **27.** $\pm\sqrt{\dfrac{441}{1024}}$ **28.** $\sqrt{\dfrac{529}{256}}$

29. $-\sqrt{\dfrac{169}{100}}$ **30.** $-\sqrt{\dfrac{289}{729}}$ **31.** $\sqrt{\dfrac{361}{2500}}$ **32.** $\pm\sqrt{\dfrac{1156}{225}}$

Example 4 $\sqrt{0.64} = \sqrt{\dfrac{64}{100}} = \dfrac{\sqrt{64}}{\sqrt{100}} = \dfrac{8}{10} = 0.8$

Find the indicated square root.

33. $\sqrt{0.16}$ **34.** $\pm\sqrt{0.49}$ **35.** $-\sqrt{1.44}$ **36.** $\sqrt{2.56}$

37. $-\sqrt{2.89}$ **38.** $\sqrt{3.24}$ **39.** $\pm\sqrt{4.84}$ **40.** $\sqrt{6.25}$

Mixed Review Exercises

Express as a fraction in simplest form.

1. 0.375 **2.** -3.2 **3.** $0.\overline{2}$

4. $1.\overline{08}$ **5.** $\dfrac{1}{2}\left(\dfrac{3}{4} - \dfrac{2}{3}\right)$ **6.** $\dfrac{3}{4}\left(\dfrac{x}{2} - \dfrac{2x}{5}\right)$

Factor completely.

7. $2n^2 - 10n - 48$ **8.** $6(x - 1) + y(x - 1)$ **9.** $9k^3 - k$

10. $4w^2 - 20w + 25$ **11.** $2x^2 + 5xy - 3y^2$ **12.** $2 - 7ab + 3a^2b^2$

11–4 *Irrational Square Roots*

Objective: To simplify radicals and to find decimal approximations of irrational square roots.

Vocabulary

Irrational numbers Real numbers that can't be expressed in the form $\frac{a}{b}$, where a and b are integers. Their exact values can't be expressed as either terminating or repeating decimals.

Property

Property of Completeness Every decimal represents a real number, and every real number can be represented by a decimal.

Example 1 Simplify: **a.** $\sqrt{256}$ **b.** $\sqrt{50}$ **c.** $2\sqrt{80}$ **d.** $\sqrt{704}$

Solution   **a.** $\sqrt{256} = \sqrt{4 \cdot 64}$ Factor within the radical sign.

$\qquad\quad = \sqrt{4} \cdot \sqrt{64}$ Use the product property of square roots.

$\qquad\quad = 2 \cdot 8$ Simplify.

$\qquad\quad = 16$

b. $\sqrt{50} = \sqrt{25 \cdot 2}$

$\qquad = \sqrt{25} \cdot \sqrt{2}$

$\qquad = 5\sqrt{2}$

c. $2\sqrt{80} = 2\sqrt{16 \cdot 5}$

$\qquad\quad = 2 \cdot 4\sqrt{5}$

$\qquad\quad = 8\sqrt{5}$

d. $\sqrt{704} = \sqrt{64 \cdot 11}$

$\qquad\quad = 8\sqrt{11}$

Simplify.

1. $\sqrt{27}$ 2. $\sqrt{20}$ 3. $\sqrt{72}$ 4. $\sqrt{32}$ 5. $\sqrt{48}$

6. $\sqrt{45}$ 7. $\sqrt{196}$ 8. $\sqrt{80}$ 9. $2\sqrt{63}$ 10. $4\sqrt{98}$

11. $7\sqrt{28}$ 12. $4\sqrt{40}$ 13. $\sqrt{441}$ 14. $\sqrt{289}$ 15. $3\sqrt{50}$

16. $12\sqrt{50}$ 17. $\sqrt{729}$ 18. $\sqrt{432}$ 19. $8\sqrt{75}$ 20. $2\sqrt{90}$

21. $\sqrt{147}$ 22. $\sqrt{288}$ 23. $\sqrt{4225}$ 24. $5\sqrt{800}$ 25. $5\sqrt{1025}$

1–4   *Irrational Square Roots* (continued)

Example 2   Approximate $\sqrt{396}$ to the nearest hundredth. Use your calculator or the table at the back of your textbook.

Solution $\sqrt{396} = \sqrt{2^2 \cdot 3^2 \cdot 11}$
$= \sqrt{2^2} \cdot \sqrt{3^2} \cdot \sqrt{11}$
$= 6\sqrt{11}$

From the table: $\sqrt{11} \approx 3.317$
$6\sqrt{11} \approx 6(3.317) \approx 19.902$

Therefore $\sqrt{396} \approx 19.90$.

Example 3   Approximate $\sqrt{0.6}$ to the nearest hundredth. Use your calculator or the table at the back of your textbook.

Solution $\sqrt{0.6} = \dfrac{\sqrt{60}}{\sqrt{100}} = \dfrac{\sqrt{60}}{10} \approx \dfrac{7.746}{10} = 0.7746$

Therefore $\sqrt{0.6} \approx 0.77$.

Exercises 26–37, use your calculator or the table at the back of the book.
pproximate each square root to the nearest tenth.

. $\sqrt{600}$	**27.** $\sqrt{200}$	**28.** $-\sqrt{800}$	**29.** $-\sqrt{500}$
. $-\sqrt{2700}$	**31.** $-\sqrt{2200}$	**32.** $\pm\sqrt{6600}$	**33.** $\pm\sqrt{4800}$

pproximate each square root to the nearest hundredth.

. $\sqrt{56}$	**35.** $\sqrt{32}$	**36.** $-\sqrt{0.7}$	**37.** $-\sqrt{0.2}$

ixed Review Exercises

nd the indicated square roots.

. $\sqrt{100}$	**2.** $-\sqrt{144}$	**3.** $\sqrt{\dfrac{9}{25}}$
$-\sqrt{\dfrac{36}{121}}$	**5.** $\sqrt{154^2}$	**6.** $\sqrt{\left(\dfrac{2}{5}\right)^2}$

mplify.

. $(13x)^2$	**8.** $(2y^3z^6)^2$	**9.** $(x + 2y)^2$
. $[10(a + 1)]^2$	**11.** $(9a^3b^7c)^2$	**12.** $(4z^2 + 3y^3)(4z^2 - 3y^3)$

NAME _____ _____ DATE _____

11–5 *Square Roots of Variable Expressions*

Objective: To find square roots of variable expressions and to use them to solve equations and problems.

Property

Property of Square Roots of Equal Numbers For any real numbers r and s:
$r^2 = s^2$ if and only if $r = s$ or $r = -s$.

CAUTION When you are finding the principal square root of a variable expression, you must be careful to use absolute value signs when needed to ensure that your answer is positive. For example, $\sqrt{x^2} = |x|$, not x.

Example 1 Simplify: **a.** $\sqrt{144x^2}$ **b.** $\sqrt{25n^8}$ **c.** $\sqrt{12a^3}$

Solution **a.** $\sqrt{144x^2} = \sqrt{144} \cdot \sqrt{x^2}$
$= 12|x|$

b. $\sqrt{25n^8} = \sqrt{25} \cdot \sqrt{n^8}$
$= \sqrt{25} \cdot \sqrt{(n^4)^2}$
$= 5n^4$ (n^4 is always nonnegative)

c. $\sqrt{12a^3} = \sqrt{4 \cdot 3 \cdot a^2 \cdot a}$
$= \sqrt{4} \cdot \sqrt{a^2} \cdot \sqrt{3} \cdot \sqrt{a}$
$= 2|a|\sqrt{3a}$

Simplify.

1. $\sqrt{81x^2}$

2. $\sqrt{121x^2}$

3. $\sqrt{20x^2}$

4. $\sqrt{45x^4}$

5. $-\sqrt{25x^2}$

6. $-\sqrt{16c^4}$

7. $-\sqrt{64d^8}$

8. $-\sqrt{98n^6}$

9. $\sqrt{225y^4}$

10. $\sqrt{400a^6b^4}$

11. $\sqrt{81m^{12}}$

12. $\sqrt{441n^6}$

13. $\pm\sqrt{75x^2y^3}$

14. $\pm\sqrt{60x^6y^4}$

15. $-\sqrt{121x^2y^2}$

16. $-\sqrt{900a^4b^6}$

17. $\pm\sqrt{\dfrac{81x^8}{100}}$

18. $\pm\sqrt{\dfrac{121}{225x^{10}}}$

19. $\sqrt{\dfrac{x^4y^8}{9z^2}}$

20. $\sqrt{\dfrac{32m^3n^2}{2mn^2}}$

21. $\sqrt{\dfrac{16x^{18}}{3600y^{20}}}$

22. $\sqrt{\dfrac{256x^{40}}{25}}$

23. $\sqrt{2.25x^4}$

24. $-\sqrt{2.56k^2}$

1-5 Square Roots of Variable Expressions (continued)

Example 2   Simplify $\sqrt{m^2 - 8m + 16}$.

Solution $\sqrt{m^2 - 8m + 16} = \sqrt{(m-4)^2} = |m-4|$

mplify.

5. $\sqrt{x^2 + 4x + 4}$

26. $\sqrt{n^2 - 14n + 49}$

7. $\sqrt{x^2 - 6x + 9}$

28. $\sqrt{m^2 - 10m + 25}$

Example 3   Solve $4x^2 = 25$.

Solution 1

$$4x^2 = 25$$
$$4x^2 - 25 = 0$$
$$(2x + 5)(2x - 5) = 0$$
$$2x = -5 \quad \text{or} \quad 2x = 5$$
$$x = -\frac{5}{2} \quad \text{or} \quad x = \frac{5}{2}$$

Check: $4\left(\frac{5}{2}\right)^2 \overset{?}{=} 25$ and $4\left(-\frac{5}{2}\right)^2 \overset{?}{=} 25$

$25 = 25 \checkmark$ and $25 = 25 \checkmark$

The solution set is $\left\{\frac{5}{2}, -\frac{5}{2}\right\}$.

Solution 2 $4x^2 = 25$

$$x^2 = \frac{25}{4}$$

$$x = \pm\sqrt{\frac{25}{4}}$$

$$x = \pm\frac{5}{2}$$

lve.

$x^2 = 16$ **30.** $n^2 = 36$ **31.** $x^2 - 9 = 0$ **32.** $d^2 - 25 = 0$

$0 = a^2 - 49$ **34.** $0 = m^2 - 64$ **35.** $2m^2 - 18 = 0$ **36.** $40b^2 - 160 = 0$

$36y^2 - 16 = 0$ **38.** $4c^2 - 25 = 0$ **39.** $0 = 49z^2 - 9$ **40.** $0 = 45x^2 - 125$

ixed Review Exercises

mplify.

1. $\pm\sqrt{80}$ **2.** $-4\sqrt{75}$ **3.** $3\sqrt{256}$

4. $2^{-3} - 3^{-2}$ **5.** $4^3 \cdot 2^{-5}$ **6.** $(3x^2)^3(-2x^4)^2$

aluate if $x = 9$, $y = 16$, and $n = 1$.

7. $x^2 + y^2$ **8.** x^2n^2 **9.** $y^2 - x^2$ **10.** $\sqrt{\dfrac{y}{n}}$ **11.** $\sqrt{\dfrac{x}{y}}$ **12.** $\left(\sqrt{y}\right)^2$

11-6 *The Pythagorean Theorem*

Objective: To use the Pythagorean theorem and its converse to solve geometric problems.

Theorems

Pythagorean Theorem In any right triangle, the square of the length of the hypotenuse equals the sum of the squares of the lengths of the legs.

$$a^2 + b^2 = c^2$$

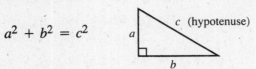

Converse of the Pythagorean Theorem If the sum of the squares of the lengths of the two shorter sides of a triangle is equal to the square of the length of the longest, then the triangle is a right triangle. The right angle is opposite the longest side.

Example 1 The length of one side of a right triangle is 24 cm.
The length of the hypotenuse is 25 cm.
Write and solve an equation to find the unknown side.

Solution
$$a^2 + b^2 = c^2$$
$$a^2 = c^2 - b^2$$
$$a = \sqrt{c^2 - b^2}$$
$$= \sqrt{25^2 - 24^2}$$
$$= \sqrt{625 - 576}$$
$$= \sqrt{49}$$
$$= 7$$

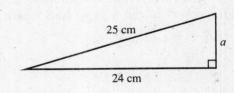

Check: $7^2 + 24^2 = 25^2$
$49 + 576 = 625 \checkmark$

The length of the third side of the right triangle is 7 cm.

In Exercises 1–10, refer to the triangle at the right.
Find the missing length correct to the nearest hundredth.
A calculator may be helpful.

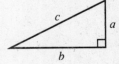

1. $a = 12$, $b = 16$, $c = $ __?__.

2. $a = 14$, $b = 48$, $c = $ __?__.

3. $a = 7$, $b = 4$, $c = $ __?__.

4. $a = 12$, $b = 8$, $c = $ __?__.

5. $a = 10$, $b = 10$, $c = $ __?__.

6. $a = 14$, $b = 7$, $c = $ __?__.

7. $a = $ __?__, $b = 10$, $c = 16$.

8. $a = $ __?__, $b = 35$, $c = 37$.

9. $a = 10$, $b = $ __?__, $c = 26$.

10. $a = 40$, $b = $ __?__, $c = 41$.

1–6 The Pythagorean Theorem *(continued)*

Example 2 State whether or not the three given numbers represent the lengths of the sides of a right triangle.

a. 5, 12, 13   **b.** 12, 18, 22

Solution **a.** $a^2 + b^2 = c^2$ **b.** $a^2 + b^2 = c^2$

$5^2 + 12^2 \overset{?}{=} 13^2$ $12^2 + 18^2 \overset{?}{=} 22^2$

$25 + 144 \overset{?}{=} 169$ $144 + 324 \overset{?}{=} 484$

$169 = 169 \ \checkmark$ $468 \neq 484$

5, 12, and 13 form   12, 18, and 22 do not
a right triangle.   form a right triangle.

ate whether or not the three given numbers represent the lengths of
e sides of a right triangle.

. 9, 16, 20   **12.** 9, 40, 41   **13.** 9, 12, 15

. 5, 6, 7   **15.** 6, 8, 10   **16.** 11, 12, 16

. 18, 24, 30   **18.** 45, 60, 75   **19.** 17, 18, 33

. 20, 21, 29   **21.** 21, 28, 35   **22.** 15, 36, 38

**Exercises 23–30, refer to the diagram for Exercises 1–10. Find the
ssing length correct to the nearest hundredth.**

. $a = b = 16$, $c = \underline{\ ?\ }$.   **24.** $a = b = 9$, $c = \underline{\ ?\ }$.

. $a = 16$, $b = \frac{1}{2}a$, $c = \underline{\ ?\ }$.   **26.** $a = \frac{1}{2}b$, $b = 10$, $c = \underline{\ ?\ }$.

. $a = 12$, $b = \frac{1}{3}a$, $c = \underline{\ ?\ }$.   **28.** $a = 16$, $b = \frac{1}{4}a$, $c = \underline{\ ?\ }$.

. $a = b = 10$, $c = \underline{\ ?\ }$.   **30.** $a = \frac{1}{2}b$, $b = 8$, $c = \underline{\ ?\ }$.

ixed Review Exercises

nplify.

$\sqrt{25x^{12}y^2}$ **2.** $\sqrt{c^2 - 8c + 16}$ **3.** $\sqrt{48a^5(b + 2)^2}$

ite a fraction in simplest form.

$(2 \cdot 10^{-2})^3$ **5.** $(3x^{-2}y^{-3})^2$ **6.** $\frac{5}{6} \div \frac{6}{5}$

$\frac{a-2}{4} + \frac{2-3a}{6}$ **8.** $4y + \frac{y-1}{y-2}$ **9.** $\frac{3r^2}{5x} \div 15rx$

$\left(-\frac{k^2}{4}\right)^3$ **11.** $\frac{2x^2 - 16x - 40}{2x - 20}$ **12.** $\frac{2x^2 + 5x - 3}{3 + x}$

11–7 Multiplying, Dividing, and Simplifying Radicals

Objective: To simplify products and quotients of radicals.

Vocabulary

Rationalize the denominator The process of eliminating a radical from the denominator of a fraction. Remember that $(\sqrt{a})^2 = a$.

Simplest form of a square-root radical

When all of the following are true:	Simplest form	*Not* in simplest form.
1. No integral radicand has a perfect-square factor other than 1.	$2\sqrt{5}$	$\sqrt{20}$
2. No fractions are under a radical sign.	$\dfrac{\sqrt{3}}{3}$	$\sqrt{\dfrac{1}{3}}$
3. No radicals are in a denominator.	$\dfrac{5\sqrt{2}}{2}$	$\dfrac{5}{\sqrt{2}}$

Example 1   Simplify $2\sqrt{3} \cdot 3\sqrt{48}$.

Solution
$$2\sqrt{3} \cdot 3\sqrt{48} = (2 \cdot 3)(\sqrt{3} \cdot \sqrt{48})$$
$$= 6\sqrt{144}$$
$$= 6 \cdot 12$$
$$= 72$$

Simplify.

1. $6\sqrt{2} \cdot 3\sqrt{2}$
2. $3\sqrt{5} \cdot 2\sqrt{5}$
3. $\sqrt{2} \cdot \sqrt{2} \cdot \sqrt{9}$
4. $\sqrt{3} \cdot \sqrt{3} \cdot \sqrt{16}$
5. $2\sqrt{3} \cdot \sqrt{5}$
6. $4\sqrt{2} \cdot \sqrt{3}$
7. $\sqrt{2} \cdot \sqrt{32}$
8. $\sqrt{3} \cdot \sqrt{27}$
9. $\sqrt{11} \cdot \sqrt{99}$
10. $\sqrt{8} \cdot \sqrt{18}$
11. $4\sqrt{108}$
12. $7\sqrt{80}$

Example 2   Simplify $\sqrt{\dfrac{7}{6}} \cdot \sqrt{\dfrac{54}{28}}$.

Solution $\sqrt{\dfrac{7}{6}} \cdot \sqrt{\dfrac{54}{28}} = \sqrt{\dfrac{7}{6} \cdot \dfrac{54}{28}} = \sqrt{\dfrac{9}{4}} = \dfrac{3}{2}$

Simplify.

13. $\sqrt{\dfrac{7}{10}} \cdot \sqrt{\dfrac{10}{7}}$
14. $\sqrt{\dfrac{5}{3}} \cdot \sqrt{\dfrac{3}{20}}$
15. $\sqrt{\dfrac{24}{11}} \cdot \sqrt{\dfrac{33}{2}}$
16. $\sqrt{\dfrac{7}{5}} \cdot \sqrt{\dfrac{5}{28}}$

17. $\sqrt{\dfrac{3}{8}} \cdot \sqrt{\dfrac{8}{27}}$
18. $\sqrt{\dfrac{3}{5}} \cdot \sqrt{\dfrac{125}{3}}$
19. $\sqrt{\dfrac{7}{3}} \cdot \sqrt{\dfrac{3}{112}}$
20. $\sqrt{\dfrac{2}{5}} \cdot \sqrt{\dfrac{10}{8}}$

11–7 Multiplying, Dividing, and Simplifying Radicals (continued)

Example 3 Simplify: **a.** $\dfrac{2}{\sqrt{3}}$ **b.** $\sqrt{\dfrac{5}{8}}$ **c.** $\dfrac{5\sqrt{2}}{\sqrt{12}}$ **d.** $\sqrt{4\dfrac{4}{5}}\cdot\sqrt{3\dfrac{1}{3}}$

Solution   **a.** $\dfrac{2}{\sqrt{3}}=\dfrac{2}{\sqrt{3}}\cdot\dfrac{\sqrt{3}}{\sqrt{3}}=\dfrac{2\sqrt{3}}{\left(\sqrt{3}\right)^2}=\dfrac{2\sqrt{3}}{3}$

b. $\sqrt{\dfrac{5}{8}}=\dfrac{\sqrt{5}}{\sqrt{8}}=\dfrac{\sqrt{5}}{2\sqrt{2}}\cdot\dfrac{\sqrt{2}}{\sqrt{2}}=\dfrac{\sqrt{5\cdot 2}}{2\left(\sqrt{2}\right)^2}=\dfrac{\sqrt{10}}{4}$

c. $\dfrac{5\sqrt{2}}{\sqrt{12}}=\dfrac{5\sqrt{2}}{\sqrt{2^2\cdot 3}}=\dfrac{5\sqrt{2}}{2\sqrt{3}}\cdot\dfrac{\sqrt{3}}{\sqrt{3}}=\dfrac{5\sqrt{6}}{2\left(\sqrt{3}\right)^2}=\dfrac{5\sqrt{6}}{6}$

d. $\sqrt{4\dfrac{4}{5}}\cdot\sqrt{3\dfrac{1}{3}}=\sqrt{\dfrac{24}{5}}\cdot\sqrt{\dfrac{10}{3}}=\sqrt{\dfrac{24}{5}\cdot\dfrac{10}{3}}=\sqrt{16}=4$

Simplify.

21. $\dfrac{3}{\sqrt{5}}$ **22.** $\dfrac{4}{\sqrt{6}}$ **23.** $\sqrt{\dfrac{1}{6}}$ **24.** $\sqrt{\dfrac{3}{8}}$

25. $\dfrac{6\sqrt{5}}{\sqrt{80}}$ **26.** $\dfrac{2\sqrt{3}}{\sqrt{48}}$ **27.** $\sqrt{3\dfrac{3}{4}}\cdot\sqrt{2\dfrac{2}{3}}$ **28.** $\sqrt{1\dfrac{1}{6}}\cdot\sqrt{4\dfrac{2}{3}}$

Example 4 Simplify $\sqrt{3}(\sqrt{3}-4)$.

Solution $\sqrt{3}(\sqrt{3}-4)=\sqrt{3}\cdot\sqrt{3}-\sqrt{3}\cdot 4$
$$=3-4\sqrt{3}$$

Simplify.

29. $\sqrt{2}(\sqrt{2}-1)$ **30.** $\sqrt{6}(5-\sqrt{6})$ **31.** $2\sqrt{3}(\sqrt{27}-\sqrt{3})$ **32.** $3\sqrt{5}(2\sqrt{5}-\sqrt{125})$

Mixed Review Exercises

Solve.

1. $x^2=121$ **2.** $2x^2-128=0$ **3.** $16x^2-1=24$

4. $\dfrac{1}{c}+\dfrac{1}{3}=\dfrac{1}{2}$ **5.** $\dfrac{3}{5}=\dfrac{15}{y}$ **6.** $\dfrac{6b-1}{3b-1}=\dfrac{5}{2}$

Simplify.

7. $15x+2(3x-5)+2$ **8.** $10a+6-(6a-12)$ **9.** $3(2a-5)-4(a-3)$

10. $(-4cd^2)(-3c^2d)$ **11.** $-3m+2+9m-5$ **12.** $x(x-1)+(x-3)(2x-1)$

11–8 Adding and Subtracting Radicals

Objective: To simplify sums and differences of radicals.

Example 1 Simplify: **a.** $3\sqrt{5} + 4\sqrt{5}$ **b.** $2\sqrt{7} + 3\sqrt{5}$

Solution **a.** Terms with common factors can be combined.

$3\sqrt{5} + 4\sqrt{5} = (3 + 4)\sqrt{5} = 7\sqrt{5}$

b. Terms that have unlike radicals cannot be combined.

Therefore $2\sqrt{7} + 3\sqrt{5}$ is in simplest form.

Simplify.

1. $6\sqrt{2} - 3\sqrt{2}$

2. $8\sqrt{3} + 4\sqrt{3}$

3. $2\sqrt{15} - \sqrt{13}$

4. $-5\sqrt{2} + 8\sqrt{2}$

5. $4\sqrt{5} + 3\sqrt{7}$

6. $9\sqrt{3} - 4\sqrt{3}$

7. $7\sqrt{5} + 6\sqrt{5}$

8. $8\sqrt{3} - \sqrt{3}$

9. $-3\sqrt{10} + 7\sqrt{6}$

Example 2 Simplify $4\sqrt{3} - 2\sqrt{7} + 8\sqrt{3}$.

Solution Use the distributive property to regroup.

$4\sqrt{3} - 2\sqrt{7} + 8\sqrt{3} = (4 + 8)\sqrt{3} - 2\sqrt{7}$

$= 12\sqrt{3} - 2\sqrt{7}$

Simplify.

10. $10\sqrt{6} - 3\sqrt{6} + \sqrt{6}$

11. $7\sqrt{2} + 6\sqrt{2} - 3\sqrt{2}$

12. $15\sqrt{7} + 2\sqrt{7} - 10\sqrt{7}$

13. $2\sqrt{3} + 5\sqrt{2} + 8\sqrt{3}$

14. $8\sqrt{6} - \sqrt{3} + \sqrt{6} - 3\sqrt{3}$

15. $2\sqrt{5} - 5\sqrt{2} + 9\sqrt{2} - 6\sqrt{5}$

16. $2\sqrt{6} - \sqrt{10} + 4\sqrt{6}$

17. $3\sqrt{5} + \sqrt{11} - 4\sqrt{11} + \sqrt{5}$

18. $6\sqrt{3} - \sqrt{7} + 8\sqrt{3} - 6\sqrt{7}$

19. $5\sqrt{13} - 3\sqrt{2} + 2\sqrt{13} - 8\sqrt{2}$

Example 3 Simplify $5\sqrt{2} - 3\sqrt{6} + 2\sqrt{8} - 5\sqrt{54}$.

Solution First express each radical in simplest form.

$5\sqrt{2} - 3\sqrt{6} + 2\sqrt{8} - 5\sqrt{54} = 5\sqrt{2} - 3\sqrt{6} + 2\sqrt{4 \cdot 2} - 5\sqrt{9 \cdot 6}$

$= 5\sqrt{2} - 3\sqrt{6} + 2(2\sqrt{2}) - 5(3\sqrt{6})$

$= 5\sqrt{2} - 3\sqrt{6} + 4\sqrt{2} - 15\sqrt{6}$

$= 9\sqrt{2} - 18\sqrt{6}$

1–8 Adding and Subtracting Radicals (continued)

Simplify.

20. $3\sqrt{2} + 3\sqrt{50}$

21. $3\sqrt{24} - 2\sqrt{6}$

22. $5\sqrt{63} - 2\sqrt{28}$

23. $5\sqrt{18} - 7\sqrt{72}$

24. $2\sqrt{80} - 6\sqrt{45}$

25. $5\sqrt{48} - 6\sqrt{27}$

26. $4\sqrt{2} + \sqrt{72}$

27. $\sqrt{108} - \sqrt{27}$

28. $3\sqrt{8} - 2\sqrt{50}$

29. $5\sqrt{12} - 3\sqrt{48}$

30. $\sqrt{32} - \sqrt{50}$

31. $\sqrt{98} - \sqrt{72}$

32. $4\sqrt{72} - 6\sqrt{32}$

33. $2\sqrt{75} + 3\sqrt{108}$

34. $10\sqrt{18} - 5\sqrt{32}$

35. $2\sqrt{2} + 3\sqrt{8} + \sqrt{32}$

36. $5\sqrt{28} + 3\sqrt{112}$

37. $4\sqrt{54} - 2\sqrt{6}$

38. $8\sqrt{8} - 4\sqrt{32} + 3\sqrt{2}$

39. $2\sqrt{27} - \sqrt{75} - \sqrt{3}$

40. $4\sqrt{75} - 3\sqrt{48} - \sqrt{27}$

41. $2\sqrt{45} - 5\sqrt{20} + 2\sqrt{5}$

Mixed Review Exercises

Write each equation in slope-intercept form.

1. $2y = 4x + 6$

2. $3y - x + 9 = 0$

3. $2x - y = 2$

4. $3x + 3y = 2$

5. $x = 2y + 10$

6. $2x - 5y = 0$

7. $x = -y + 7$

8. $4 - x + 2y = 0$

9. $6x - 9 = 3y$

10. $8y + 12 = 4x$

For each parabola whose equation is given, find the coordinates of the vertex and the equation of the axis of symmetry.

11. $y = -2x^2$

12. $y = x^2 - 4x + 4$

13. $y = 3 - 2x^2$

14. $y = -x^2 - 4x$

Solve. Assume that no denominator equals zero.

15. $\dfrac{x - 3}{8} = \dfrac{x}{5}$

16. $\dfrac{2}{x + 4} = \dfrac{-1}{2x + 3}$

17. $9 - \dfrac{8}{x} = x$

18. $\dfrac{1}{x} = \dfrac{2x}{x + 1}$

19. $\dfrac{2}{x + 1} + \dfrac{1}{x - 1} = 1$

20. $\dfrac{x + 1}{x + 4} - \dfrac{1}{8} = \dfrac{x - 3}{x}$

11–9 Multiplication of Binomials Containing Radicals

Objective: To multiply binomials containing square-root radicals and to rationalize binomial denominators that contain square-root radicals.

Vocabulary

Conjugates If b and d are both nonnegative, then the binomials

$$a\sqrt{b} + c\sqrt{d} \quad \text{and} \quad a\sqrt{b} - c\sqrt{d}$$

are called conjugates of one another.

Example 1 Simplify $(4 + \sqrt{5})(4 - \sqrt{5})$.

Solution The pattern is $(a + b)(a - b) = a^2 - b^2$.

$$(4 + \sqrt{5})(4 - \sqrt{5}) = 4^2 - (\sqrt{5})^2$$
$$= 16 - 5$$
$$= 11$$

Example 2 Simplify $(2 + \sqrt{3})^2$.

Solution The pattern is $(a + b)^2 = a^2 + 2ab + b^2$.

$$(2 + \sqrt{3})^2 = 2^2 + 2(2)(\sqrt{3}) + (\sqrt{3})^2$$
$$= 4 + 4\sqrt{3} + 3$$
$$= 7 + 4\sqrt{3}$$

Simplify.

1. $(5 + \sqrt{3})(5 - \sqrt{3})$
2. $(2 - \sqrt{3})(2 + \sqrt{3})$
3. $(3 + \sqrt{7})(3 - \sqrt{7})$
4. $(\sqrt{15} - 2)(\sqrt{15} + 2)$
5. $(\sqrt{11} - 5)(\sqrt{11} + 5)$
6. $(\sqrt{7} + 6)(\sqrt{7} - 6)$
7. $(\sqrt{3} + 5)(\sqrt{3} - 5)$
8. $(6 - \sqrt{5})(6 + \sqrt{5})$
9. $(2\sqrt{3} - 1)^2$
10. $(4\sqrt{6} + 3)^2$
11. $(8 + \sqrt{3})^2$
12. $(\sqrt{5} + 3)^2$

Example 3 Simplify $(3\sqrt{2} - 7\sqrt{5})^2$.

Solution The pattern is $(a - b)^2 = a^2 - 2ab + b^2$.

$$(3\sqrt{2} - 7\sqrt{5})^2 = (3\sqrt{2})^2 - 2(3\sqrt{2})(7\sqrt{5}) + (7\sqrt{5})^2$$
$$= (3\sqrt{2})^2 - 2(3)(7)(\sqrt{2})(\sqrt{5}) + (7\sqrt{5})^2$$
$$= 9(2) - 42\sqrt{10} + 49(5)$$
$$= 18 - 42\sqrt{10} + 245$$
$$= 263 - 42\sqrt{10}$$

Study Guide, ALGEBRA, Structure and Method, Book 1

1–9 Multiplication of Binomials Containing Radicals (continued)

implify.

3. $(2\sqrt{5} - \sqrt{3})(2\sqrt{5} + \sqrt{3})$

14. $(3\sqrt{7} - \sqrt{5})(3\sqrt{7} + \sqrt{5})$

. $(4\sqrt{5} - \sqrt{17})(4\sqrt{5} + \sqrt{17})$

16. $(6\sqrt{10} + 2\sqrt{6})(6\sqrt{10} - 2\sqrt{6})$

7. $(2\sqrt{3} - 3)(3\sqrt{3} + 2)$

18. $(5\sqrt{2} + 2)(3\sqrt{2} - 2)$

. $(2\sqrt{5} - 5\sqrt{7})(3\sqrt{5} + \sqrt{7})$

20. $(6\sqrt{11} + \sqrt{6})(2\sqrt{11} + 3\sqrt{6})$

.. $(5\sqrt{11} - 2\sqrt{3})(4\sqrt{11} + 3\sqrt{3})$

22. $(4\sqrt{6} - 3\sqrt{2})(6\sqrt{6} - 2\sqrt{2})$

Example 4 Rationalize the denominator of $\dfrac{2}{3 + \sqrt{5}}$.

Solution $\dfrac{2}{3 + \sqrt{5}} = \dfrac{2}{3 + \sqrt{5}} \cdot \dfrac{3 - \sqrt{5}}{3 - \sqrt{5}}$ Use the conjugate of the denominator.

$$= \dfrac{2(3 - \sqrt{5})}{9 - (\sqrt{5})^2}$$

$$= \dfrac{6 - 2\sqrt{5}}{9 - 5}$$

$$= \dfrac{6 - 2\sqrt{5}}{4}$$

---

tionalize the denominator of each fraction.

. $\dfrac{1}{1 + \sqrt{3}}$

24. $\dfrac{1}{3 + \sqrt{2}}$

25. $\dfrac{2}{\sqrt{5} - 1}$

26. $\dfrac{1}{\sqrt{7} - 2}$

. $\dfrac{2 + \sqrt{5}}{1 - \sqrt{5}}$

28. $\dfrac{\sqrt{5} - 1}{\sqrt{5} + 2}$

29. $\dfrac{6}{2\sqrt{3} - 1}$

30. $\dfrac{7}{3\sqrt{7} + 2}$

ixed Review Exercises

mplify. Assume that all variables represent positive real numbers.

. $\sqrt{72x^6}$

2. $3\sqrt{12x} \cdot 4\sqrt{3}$

3. $4\sqrt{20} - 2\sqrt{5}$

. $3\sqrt{27} + 5\sqrt{48}$

5. $\sqrt{\dfrac{2}{3}} \cdot \sqrt{\dfrac{3}{32}}$

6. $\sqrt{1\dfrac{2}{3}} \cdot \sqrt{3\dfrac{1}{3}}$

. $(2 - 3k^2)^2$

8. $(3p + 5z)^2$

9. $(4ab + x)(4ab - x)$

ve.

. $5p - 1 = 4(p + 3)$

11. $x^2 - 11x + 28 = 0$

12. $4g^2 = 100$

11–10 Simple Radical Equations

Objective: To solve simple radical equations.

Vocabulary

> **Radical equation** An equation containing a radical with a variable in the radicand.
> For example, $\sqrt{5x + 9} = 12$.

CAUTION   When you square both sides of an equation, the new equation
may not be equivalent to the original equation. Therefore,
you must *check every possible root in the original equation* to
see whether it is indeed a root.

Example 1   Solve $\sqrt{3x + 1} = 5$.

Solution

$$\sqrt{3x + 1} = 5$$
$$(\sqrt{3x + 1})^2 = 5^2 \quad \left\{ \begin{array}{l} \text{Square both sides} \\ \text{of the equation.} \end{array} \right.$$
$$3x + 1 = 25$$
$$3x = 24$$
$$x = 8$$

The solution set is $\{8\}$.

Check:
$$\sqrt{3(8) + 1} \overset{?}{=} 5$$
$$\sqrt{24 + 1} \overset{?}{=} 5$$
$$\sqrt{25} \overset{?}{=} 5$$
$$5 = 5 \ \checkmark$$

Solve.

1. $\sqrt{x} = 4$

2. $\sqrt{y} = 9$

3. $\sqrt{x} = 12$

4. $\sqrt{x} = 15$

5. $\sqrt{d} = 20$

6. $\sqrt{e} = 25$

7. $\sqrt{2x} = 8$

8. $12 = \sqrt{3a}$

9. $4 = \sqrt{\dfrac{x}{3}}$

10. $6 = \sqrt{\dfrac{x}{2}}$

11. $\sqrt{x - 1} = 7$

12. $\sqrt{x + 2} = 6$

13. $\sqrt{x + 1} = 4$

14. $\sqrt{m + 4} = 1$

15. $15 = 5\sqrt{3x}$

16. $\sqrt{3x - 2} = 7$

17. $\sqrt{7x + 1} = 8$

18. $\sqrt{2x + 5} = 7$

19. $\sqrt{x} = 2\sqrt{5}$

20. $\sqrt{y} = 3\sqrt{2}$

21. $\sqrt{x} = 6\sqrt{3}$

Example 2   Solve $\sqrt{4x + 1} + 3 = 8$.

Solution

$$\sqrt{4x + 1} + 3 = 8 \quad \left\{ \begin{array}{l} \text{First set apart the radical on} \\ \text{one side of the equals sign.} \end{array} \right.$$
$$\sqrt{4x + 1} = 5$$
$$4x + 1 = 25 \quad \text{Then square each side.}$$
$$4x = 24$$
$$x = 6$$

The solution set is $\{6\}$.

Check:
$$\sqrt{4(6) + 1} + 3 \overset{?}{=} 8$$
$$\sqrt{24 + 1} + 3 \overset{?}{=} 8$$
$$\sqrt{25} + 3 \overset{?}{=} 8$$
$$5 + 3 \overset{?}{=} 8$$
$$8 = 8 \ \checkmark$$

1–10 Simple Radical Equations (continued)

Solve.

22. $1 = \sqrt{m} - 2$

23. $6 = \sqrt{x} - 2$

24. $\sqrt{a} - 5 = 4$

25. $\sqrt{2m - 1} + 2 = 9$

26. $\sqrt{4x + 1} - 1 = 2$

27. $\sqrt{2x + 3} + 5 = 8$

28. $\sqrt{3m - 2} - 4 = 1$

29. $10 = 7 + \sqrt{2m - 1}$

30. $\sqrt{5y + 1} + 5 = 11$

Example 3 Solve $\sqrt{5m^2 - 36} = 2m$.

Solution
$$\sqrt{5m^2 - 36} = 2m$$
$$5m^2 - 36 = 4m^2 \quad \text{Square each side of the equation.}$$
$$m^2 = 36$$
$$m = 6 \quad \text{or} \quad m = -6$$

Check:
$$\sqrt{5(6)^2 - 36} \stackrel{?}{=} 2(6) \qquad \sqrt{5(-6)^2 - 36} \stackrel{?}{=} 2(-6)$$
$$\sqrt{5(36) - 36} \stackrel{?}{=} 12 \qquad \sqrt{5(36) - 36} \stackrel{?}{=} -12$$
$$\sqrt{180 - 36} \stackrel{?}{=} 12 \qquad \sqrt{180 - 36} \stackrel{?}{=} -12$$
$$\sqrt{144} \stackrel{?}{=} 12 \qquad \qquad \sqrt{144} \stackrel{?}{=} -12$$
$$12 = 12 \checkmark \qquad \qquad 12 \neq -12$$

The solution set is $\{6\}$.

Solve.

31. $\sqrt{3a^2 - 2} = 5$

32. $\sqrt{2a^2 + 1} = 3$

33. $\sqrt{c^2 + 15} = 4c$

34. $\sqrt{5x^2 - 1} = x$

35. $\sqrt{\dfrac{2x + 5}{3}} = 5$

36. $\sqrt{\dfrac{2n - 4}{6}} = 2$

37. $4 = \sqrt{\dfrac{5c - 3}{7}}$

38. $4 = \sqrt{\dfrac{3k + 8}{2}}$

39. $3 = \sqrt{\dfrac{4x - 1}{7}}$

Mixed Review Exercises

Express in simplest form.

1. $(3 + \sqrt{6})(3 - \sqrt{6})$

2. $(2 + \sqrt{3})^2$

3. $\dfrac{2}{2 + \sqrt{5}}$

4. $\dfrac{2 + \sqrt{3}}{1 - \sqrt{3}}$

5. $3\sqrt{2}(\sqrt{18} - 2\sqrt{2})$

6. $(2\sqrt{5} + 1)(\sqrt{5} - 3)$

Factor completely.

7. $5a^2 - 10a + 5$

8. $t^3 - t^2 - 20t$

9. $3x(x + 1) + 2(x + 1)$

10. $x^3 + x^2 - 3x - 3$

11. $3x^5 - 48x$

12. $2x^2 - 12xy + 18y^2$

---

12 Quadratic Functions

12-1 *Quadratic Equations with Perfect Squares*

Objective: To solve quadratic equations involving perfect squares.

Vocabulary

Perfect square An expression such as x^2, $(x - 1)^2$, or $(2x + 5)^2$.

Roots of $x^2 = k$ An equation in the form "perfect square $= k$" $(k \geq 0)$ can be solved by the method shown in Examples 1 and 2.

If $k > 0$,   then $x^2 = k$ has **2** real roots: $x = \pm\sqrt{k}$.
If $k = 0$,   then $x^2 = k$ has **1** real root: $x = 0$.
If $k < 0$,   then $x^2 = k$ has **no** real roots.

Example 1   Solve: **a.** $m^2 = 36$ **b.** $3r^2 = 48$ **c.** $x^2 - 11 = 0$ **d.** $m^2 = -25$

Solution   **a.** $m^2 = 36$
  $m = \pm\sqrt{36}$
  $m = \pm 6$

  The solution set is $\{6, -6\}$.

b. $3r^2 = 48$
  $r^2 = 16$
  $r = \pm\sqrt{16} = \pm 4$

  The solution set is $\{4, -4\}$.

c. $x^2 - 11 = 0$
  $x^2 = 11$
  $x = \pm\sqrt{11}$

  The solution set is $\{\sqrt{11}, -\sqrt{11}\}$.

d. $m^2 = -25$
  Since the square of any real number is always a nonnegative number, there is *no real solution*.

Solve. Express irrational solutions in simplest radical form. If the equation has no solution, write *No solution*.

1. $x^2 = 49$ **2.** $2x^2 = 18$ **3.** $x^2 = \dfrac{25}{36}$ **4.** $a^2 = -16$

5. $2x^2 = 128$ **6.** $5x^2 = 125$ **7.** $9x^2 = 81$ **8.** $x^2 - 81 = 0$

9. $x^2 + 25 = 0$ **10.** $m^2 - 10 = 0$ **11.** $0 = 6x^2 - 24$ **12.** $0 = 3m^2 - 75$

Example 2   Solve $(x + 3)^2 = 49$

Solution $(x + 3)^2 = 49$ Note that $(x + 3)^2$ is a perfect square.
      $x + 3 = \pm\sqrt{49}$ Find the *positive or negative* square root of each side.
      $x = -3 \pm 7$ Subtract 3 from each side.
      $x = 4$ or $x = -10$

  Check: $(4 + 3)^2 \overset{?}{=} 49$ $(-10 + 3)^2 \overset{?}{=} 49$
      $7^2 \overset{?}{=} 49$ $(-7)^2 \overset{?}{=} 49$
      $49 = 49 \checkmark$ $49 = 49 \checkmark$

  The solution set is $\{4, -10\}$.

2–1 Quadratic Equations with Perfect Squares (continued)

Solve. Express irrational solutions in simplest radical form. If the
equation has no solution, write *No solution*.

13. $(x - 3)^2 = 0$ **14.** $(z - 1)^2 = 16$ **15.** $(r - 5)^2 = 100$

16. $(x - 1)^2 = 25$ **17.** $(2x + 9)^2 = 225$ **18.** $(t - 4)^2 = 9$

Example 3   Solve: **a.** $3(x - 2)^2 = 21$ **b.** $y^2 + 10y + 25 = 36$

Solution   **a.** $3(x - 2)^2 = 21$ **b.** $y^2 + 10y + 25 = 36$
$$(x - 2)^2 = 7$$
$$x - 2 = \pm\sqrt{7}$$
$$x = 2 \pm \sqrt{7}$$
$$x = 2 + \sqrt{7} \quad \text{or} \quad x = 2 - \sqrt{7}$$

$$(y + 5)^2 = 36$$
$$y + 5 = \pm\sqrt{36}$$
$$y + 5 = \pm 6$$
$$y = -5 \pm 6$$
$$y = 1 \quad \text{or} \quad y = -11$$

The check is left to you.
The solution set is
$\{2 + \sqrt{7}, 2 - \sqrt{7}\}$.

The check is left to you.
The solution set is $\{1, -11\}$.

Note: Example 3(b) could also have been solved by factoring.

Solve. Express irrational solutions in simplest radical form. If the
equation has no solution, write *No solution*.

19. $9m^2 - 1 = 35$ **20.** $27 = 2r^2 - 5$ **21.** $3x^2 - 9 = 33$

22. $64 = 2t^2 - 8$ **23.** $2n^2 + 6 = 38$ **24.** $7x^2 + 1 = 64$

25. $3(m - 2)^2 = 15$ **26.** $400 = 4(z - 2)^2$ **27.** $2(x - 5)^2 = 98$

28. $25 = (2x + 1)^2$ **29.** $5(m - 3)^2 = 80$ **30.** $6(z + 5)^2 = 216$

31. $3(x - 1)^2 = -24$ **32.** $(3x - 1)^2 + 12 = 4$ **33.** $6(x + 5)^2 = 24$

34. $7(x + 2)^2 = 112$ **35.** $(x - 2)^2 - 1 = 35$ **36.** $2(3n - 1)^2 = 8$

37. $3(2x - 1)^2 = 27$ **38.** $2(x + 3)^2 - 4 = 68$ **39.** $5(x - 1)^2 + 3 = 23$

40. $x^2 - 2x + 1 = 9$ **41.** $x^2 + 18x + 81 = 98$ **42.** $x^2 - 12x + 36 = 64$

43. $x^2 - 4x + 4 = 16$ **44.** $x^2 + 10x + 25 = 81$ **45.** $n^2 - 8n + 16 = 36$

Mixed Review Exercises

Express each square as a trinomial.

1. $(x - 8)^2$ **2.** $(2x + 1)^2$ **3.** $(4x - 3)^2$

4. $(-2c + 3)^2$ **5.** $\left(x + \frac{1}{4}\right)^2$ **6.** $\left(x - \frac{1}{5}\right)^2$

7. $\left(\frac{1}{2}x + \frac{1}{3}\right)^2$ **8.** $\left(\frac{1}{4}x - \frac{2}{3}\right)^2$ **9.** $(x + 11)^2$

12–2 *Completing The Square*

Objective: To solve quadratic equations by completing the square.

Method of Completing the Square

A method of transforming a quadratic equation so that it is in the form

$$\text{perfect square} = k \quad (k \geq 0).$$

For $x^2 + bx + \underline{?}$:

1. Find half the coefficient of x: $\qquad\qquad\qquad\qquad\qquad \dfrac{b}{2}$

2. Square the result of Step 1: $\qquad\qquad\qquad\qquad \left(\dfrac{b}{2}\right)^2$

3. Add the result of Step 2 to $x^2 + bx$: $\qquad x^2 + bx + \left(\dfrac{b}{2}\right)^2$

4. You have completed the square: $\qquad x^2 + bx + \left(\dfrac{b}{2}\right)^2 = \left(x + \dfrac{b}{2}\right)^2$

Example 1 Complete the square:

$\qquad$ **a.** $x^2 + 6x + \underline{?}$ $\qquad\qquad$ **b.** $x^2 - 4x + \underline{?}$

Solution **a.** $\underline{x^2 + 6x} + 9 = (x + 3)^2$ $\qquad$ **b.** $\underline{x^2 - 4x} + 4 = (x - 2)^2$

$\qquad\qquad\qquad \left(\dfrac{6}{2}\right)^2 = 9$ $\qquad\qquad\qquad\qquad \left(-\dfrac{4}{2}\right)^2 = 4$

Complete the square.

1. $x^2 - 8x + \underline{?}$ $\qquad\qquad$ 2. $x^2 - 10x + \underline{?}$ $\qquad\qquad$ 3. $v^2 - 20v + \underline{?}$

4. $c^2 - 12c + \underline{?}$ $\qquad\qquad$ 5. $x^2 - 18x + \underline{?}$ $\qquad\qquad$ 6. $x^2 + 2x + \underline{?}$

Example 2 Solve $x^2 + 3x - 10 = 0$ by completing the square.

Solution $\qquad\qquad\qquad x^2 + 3x = 10 \qquad\qquad$ Move the constant term to the right side.

$\qquad\qquad x^2 + 3x + \left(\dfrac{3}{2}\right)^2 = 10 + \left(\dfrac{3}{2}\right)^2 \qquad$ Half of the coefficient of x is $\dfrac{3}{2}$.

$\qquad\qquad\quad x^2 + 3x + \dfrac{9}{4} = 10 + \dfrac{9}{4} \qquad$ Square it and add the result to *both* sides.

$\qquad \text{perfect square} \longrightarrow \left(x + \dfrac{3}{2}\right)^2 = \dfrac{49}{4} \longleftarrow \text{constant}$

$$x + \dfrac{3}{2} = \pm \dfrac{7}{2}$$

$$x = -\dfrac{3}{2} \pm \dfrac{7}{2}$$

$$= \dfrac{-3 \pm 7}{2}$$

$$x = 2 \quad \text{or} \quad x = -5$$

The check is left for you. The solution set is $\{-5, 2\}$.

12–2 Completing the Square (continued)

Solve by completing the square.

7. $x^2 + 8x = -15$ **8.** $x^2 + 6x = -8$ **9.** $x^2 + 4x = 5$

10. $x^2 - 7x - 8 = 0$ **11.** $x^2 - 3x - 4 = 0$ **12.** $x^2 + x - 12 = 0$

13. $y^2 + 7y = -12$ **14.** $n^2 - 5n = 36$ **15.** $x^2 + 3x = 10$

16. $y^2 - 9y + 8 = 0$ **17.** $x^2 - 5x - 50 = 0$ **18.** $x^2 - 3x - 40 = 0$

Example 3   Solve $3x^2 - 5x - 2 = 0$ by completing the square.

Solution

$$3x^2 - 5x = 2$$

$$x^2 - \frac{5}{3}x = \frac{2}{3}$$

Move the constant term to the right side.
Divide both sides by 3 so that the coefficient of x^2 will be 1.

$$x^2 - \frac{5}{3}x + \left(\frac{5}{6}\right)^2 = \frac{2}{3} + \left(\frac{5}{6}\right)^2$$

Complete the square.

$$x^2 - \frac{5}{3}x + \frac{25}{36} = \frac{2}{3} + \frac{25}{36}$$

$$\left(x - \frac{5}{6}\right)^2 = \frac{49}{36}$$

$$x - \frac{5}{6} = \pm\frac{7}{6}$$

$$x = \frac{5}{6} \pm \frac{7}{6}$$

$$x = 2 \quad \text{or} \quad x = -\frac{1}{3}$$

The check is left for you. The solution set is $\left\{2, -\frac{1}{3}\right\}$.

Solve by completing the square.

19. $2n^2 - n - 6 = 0$ **20.** $2x^2 + x - 1 = 0$ **21.** $2x^2 - 5x = 12$

22. $2x^2 - 5x - 3 = 0$ **23.** $3x^2 + 11x = 4$ **24.** $5x^2 + 4x - 1 = 0$

Solve the equations by (a) completing the square and (b) by factoring.

25. $x^2 - 12x + 35 = 0$ **26.** $x^2 + 18x + 45 = 0$ **27.** $a^2 + 17a - 60 = 0$

28. $3x^2 - 5x = 2$ **29.** $2x^2 + x = 10$ **30.** $2x^2 - 11x = -5$

Solve. Write irrational roots in simplest radical form.

31. $y^2 + 8y = -10$ **32.** $4x^2 - 4x = 7$ **33.** $3x^2 - 6x - 2 = 0$

Mixed Review Exercises

Simplify.

1. $\sqrt{140}$ **2.** $\sqrt{1200}$ **3.** $\sqrt{16x^4y^7}$ **4.** $\sqrt{60a^6b^5c^3}$

5. $\frac{x}{3 - x} + \frac{4}{x + 3}$ **6.** $1 + \frac{n^2}{n^2 - 1}$ **7.** $2(3x - 1) + (x + 5)(3x + 1)$

12–3 The Quadratic Formula

Objective: To learn the quadratic formula and use it to solve equations.

The Quadratic Formula

The solutions of a quadratic equation in the form of $ax^2 + bx + c = 0$,
$a \neq 0$ and $b^2 - 4ac \geq 0$ are given by the formula

$$x = \frac{-b \pm \sqrt{b^2 - 4ac}}{2a}.$$

Example 1 Use the quadratic formula to solve $3x^2 + 5x - 2 = 0$.

Solution $x = \dfrac{-b \pm \sqrt{b^2 - 4ac}}{2a}$, where $a = 3$, $b = 5$, and $c = -2$.

$x = \dfrac{-5 \pm \sqrt{(5)^2 - 4(3)(-2)}}{2(3)}$ Substitute the given values of a, b, and c.

$= \dfrac{-5 \pm \sqrt{25 + 24}}{6}$

$= \dfrac{-5 \pm \sqrt{49}}{6} = \dfrac{-5 \pm 7}{6}$

$x = \dfrac{-5 + 7}{6} = \dfrac{2}{6} = \dfrac{1}{3}$ or $x = \dfrac{-5 - 7}{6} = \dfrac{-12}{6} = -2$

The check is left to you. The solution set is $\left\{\dfrac{1}{3}, -2\right\}$.

Use the quadratic formula to solve each equation.

1. $x^2 + 3x - 10 = 0$

2. $x^2 - 8x + 7 = 0$

3. $x^2 + 2x - 3 = 0$

4. $x^2 - 14x + 24 = 0$

5. $n^2 + 5n - 6 = 0$

6. $x^2 - 6x - 40 = 0$

7. $2x^2 + 3x - 2 = 0$

8. $3u^2 - 5u - 2 = 0$

9. $3x^2 - 10x - 8 = 0$

10. $3x^2 - 2x - 1 = 0$

11. $2x^2 - 5x - 7 = 0$

12. $5x^2 + 6x - 8 = 0$

Example 2 Use the quadratic formula to solve $x^2 = x - 6$.

Solution $x^2 - x + 6 = 0$ Rewrite the equation in standard form.

$x = \dfrac{-b \pm \sqrt{b^2 - 4ac}}{2a}$, where $a = 1$, $b = -1$, and $c = 6$.

$x = \dfrac{-(-1) \pm \sqrt{(-1)^2 - 4(1)(6)}}{2(1)} = \dfrac{1 \pm \sqrt{1 - 24}}{2} = \dfrac{1 \pm \sqrt{-23}}{2}$

Since $\sqrt{b^2 - 4ac} = \sqrt{-23}$ and $\sqrt{-23}$ isn't a real number, there is *no real solution*.

2–3 The Quadratic Formula (continued)

Use the quadratic formula to solve each equation.

13. $x^2 - 4x + 6 = 0$
14. $2x^2 = 3x - 1$
15. $x^2 - 4x = 30$

16. $2x^2 + 2x + 5 = 0$
17. $4x^2 + 20x = -9$
18. $3x^2 - 3x + 4 = 0$

Example 3 Use the quadratic formula to solve $2x^2 - 3x - 4 = 0$. Give irrational roots in simplest radical form and then approximate them to the nearest tenth. You may wish to use a calculator.

Solution $2x^2 - 3x - 4 = 0$

$$x = \frac{-b \pm \sqrt{b^2 - 4ac}}{2a}, \text{ where } a = 2, b = -3, \text{ and } c = -4.$$

$$x = \frac{3 \pm \sqrt{9 - 4(2)(-4)}}{2(2)} \quad \text{Substitute the given values of } a, b, \text{ and } c.$$

$$= \frac{3 \pm \sqrt{9 + 32}}{4} \quad \text{Simplify.}$$

$$= \frac{3 \pm \sqrt{41}}{4}$$

Since $\sqrt{41} \approx 6.40$, $x \approx \dfrac{3 + 6.4}{4} = 2.35 \approx 2.4$

or $x \approx \dfrac{3 - 6.4}{4} = -0.85 \approx -0.9$

The check is left to you.

The solution set is $\left\{ \dfrac{3 + \sqrt{41}}{4}, \dfrac{3 - \sqrt{41}}{4} \right\}$ or $\{2.4, -0.9\}$.

Use the quadratic formula to solve each equation. Give irrational roots in simplest radical form and then approximate them to the nearest tenth. You may wish to use a calculator.

19. $2x^2 = 8x - 5$
20. $3x^2 + 2x = 2$
21. $x^2 - 4x - 10 = 0$

22. $x^2 - 4x - 2 = 0$
23. $2x^2 - 4x + 1 = 0$
24. $3x^2 - 8x + 2 = 0$

25. $2x^2 + 1 = 3x$
26. $3x^2 + x = 2$
27. $4x^2 - 11x = 3$

Mixed Review Exercises

Solve each open sentence and graph its solution set.

1. $|x - 2| \le 5$
2. $2|y + 5| = 4$
3. $|2n + 3| < 5$

4. $1 < 2z + 1 \le 7$
5. $\sqrt{x} = 5$
6. $\sqrt{5n + 1} = 6$

7. $2\sqrt{2x} = 12$
8. $|3 + 2k| = 11$
9. $3|2 - m| = 12$

Solve by completing the square.

10. $x^2 - 8x + 12 = 0$
11. $3x^2 + 6x = 0$
12. $c^2 - c = 12$

12–4 *Graphs of Quadratic Equations: The Discriminant*

Objective: To use the discriminant to find the number of roots of the equation $ax^2 + bx + c = 0$ and the number of x-intercepts of the graph of the related equation $y = ax^2 + bx + c$.

Vocabulary

x-intercept The x-coordinate of a point where the curve of a parabola intersects the x-axis.

Discriminant For a quadratic equation in the form $ax^2 + bx + c = 0$, the value of $b^2 - 4ac$ is called the discriminant.

	Value of $b^2 - 4ac$	Number of different real roots of $ax^2 + bx + c = 0$	Number of x-intercepts of the graph of $y = ax^2 + bx + c$
Case 1	positive	2	2
Case 2	zero	1 (a double root)	1
Case 3	negative	0	0

Example 1   Write the value of the discriminant of each equation. Then use it to decide how many different real-number roots the equation has. (Do not solve.)

    a. $x^2 - 4x - 5 = 0$　　**b.** $x^2 - 6x + 9 = 0$　　**c.** $x^2 - 3x + 5 = 0$

Solution   **a.** $x^2 - 4x - 5 = 0$

Substitute $a = 1$, $b = -4$, and $c = -5$ in the discriminant formula.

$b^2 - 4ac = (-4)^2 - 4(1)(-5) = 16 - 4(-5) = 16 + 20 = 36$

Since the discriminant is positive, $x^2 - 4x - 5 = 0$ has two real roots.

b. $x^2 - 6x + 9 = 0$

Substitute $a = 1$, $b = -6$, and $c = 9$ in the discriminant formula.

$b^2 - 4ac = (-6)^2 - 4(1)(9) = 36 - 36 = 0$

Since the discriminant is zero, $x^2 - 6x + 9 = 0$ has one real root.

c. $x^2 - 3x + 5 = 0$

Substitute $a = 1$, $b = -3$, and $c = 5$ in the discriminant formula.

$b^2 - 4ac = (-3)^2 - 4(1)(5) = 9 - 20 = -11$

Since the discriminant is negative, $x^2 - 3x + 5 = 0$ has no real roots.

12–4 Graphs of Quadratic Equations: The Discriminant (continued)

Write the value of the discriminant of each equation. Then use it to decide how many different real-number roots the equation has. (Do not solve.)

1. $x^2 - 3x + 2 = 0$
2. $x^2 - 3x + 5 = 0$
3. $x^2 - 8x + 16 = 0$
4. $2x^2 - 5x - 4 = 0$
5. $4y^2 - 12y + 9 = 0$
6. $3t^2 - 5t + 3 = 0$
7. $2n^2 + n - 6 = 0$
8. $5y^2 - 8y + 4 = 0$
9. $2x^2 - 7x + 5 = 0$
10. $4m^2 - 20m + 25 = 0$
11. $3x^2 - 7x + 2 = 0$
12. $-3b^2 + 2b - 3 = 0$
13. $-2x^2 + 6x - 3 = 0$
14. $3x^2 - 4x = 6$
15. $x^2 - x + 1 = 0$

Example 2 Determine (a) how many x-intercepts the parabola $y = 4x - x^2 + 5$ has and (b) whether its vertex lies above, below, or on the x-axis. (Do not draw the graph.)

Solution a. The x-intercepts of the graph are the roots of the equation

$$0 = 4x - x^2 + 5, \quad \text{or} \quad -x^2 + 4x + 5 = 0.$$

Its discriminant is $b^2 - 4ac = (4)^2 - 4(-1)(5) = 36$, which is positive.

The equation has two real roots.
The parabola has two x-intercepts.

b. Since the coefficient of x^2 is negative, the parabola opens downward. Its vertex must be above the x-axis (otherwise, the parabola would not intersect the x-axis in two points).

Without drawing the graph of the given equation, determine (a) how many x-intercepts the parabola has and (b) whether its vertex lies above, below, or on the x-axis.

16. $y = x^2 - x - 6$
17. $y = -x^2 + 3x + 4$
18. $y = x^2 + 9 - 6x$
19. $y = 2x^2 + x + 3$
20. $y = 5x - 2 + 3x^2$
21. $y = 4x^2 + 4x + 1$

Mixed Review Exercises

Simplify. Assume no denominator equals zero.

1. $\dfrac{\sqrt{2} - 1}{\sqrt{3}}$
2. $\sqrt{\dfrac{8c^3}{3}} \cdot \sqrt{\dfrac{6c}{25}}$
3. $3\sqrt{27} - 10\sqrt{3} + \sqrt{12}$
4. $\dfrac{2x + 6}{x^3 + 5x^2 + 6x}$
5. $\dfrac{x}{x - 1} + \dfrac{3}{2x + 2}$
6. $(3.1 \cdot 10^3)(4.2 \cdot 10^2)$

Find the vertex and the axis of symmetry of the graph of each equation.

7. $y = 2x^2$
8. $y = x^2 + 6x + 9$
9. $y = 2x^2 + 5$
10. $y = x^2 - 2x$
11. $y = -x^2 + 4x$
12. $y = x^2 + 4x - 5$

12–5 Methods of Solution

Objective: To choose the best method for solving a quadratic equation.

Methods for Solving a Quadratic Equation	When to Use the Method
1. Using the quadratic formula	1. If an equation is in the form $ax^2 + bx + c = 0$, especially if you use a calculator.
2. Factoring	2. If an equation is in the form $ax^2 + bx = 0$, or if the factors are easily seen.
3. Using the property of square roots of equal numbers	3. If an equation is in the form $ax^2 + c = 0$.
4. Completing the square	4. If an equation is in the form $x^2 + bx + c = 0$ and b is an even number.

Example Solve each quadratic equation using the most appropriate method.

a. $6x^2 - 54 = 0$ **b.** $2x^2 - 7x + 5 = 0$
c. $2t^2 - 28t = 0$ **d.** $n^2 + 6n - 16 = 0$

Solution **a.** $6x^2 - 54 = 0$ The equation has the form $ax^2 + c = 0$.
$$6x^2 = 54$$ Therefore, use the property of square roots of equal numbers.
$$x^2 = 9$$
$$x = \pm 3$$

The solution set is $\{-3, 3\}$.

b. $2x^2 - 7x + 5 = 0$ The equation has the form $ax^2 + bx + c = 0$.

$$x = \frac{-(-7) \pm \sqrt{(-7)^2 - 4(2)(5)}}{2(2)}$$ Therefore, use the quadratic formula.

$$= \frac{7 \pm \sqrt{49 - 40}}{4}$$

$$= \frac{7 \pm \sqrt{9}}{4}$$

$$= \frac{7 \pm 3}{4}$$

$$x = \frac{7 + 3}{4} = \frac{10}{4} \quad \text{or} \quad x = \frac{7 - 3}{4} = \frac{4}{4}$$

The solution set is $\left\{\frac{5}{2}, 1\right\}$.

c. $2t^2 - 28t = 0$ The equation has the form $ax^2 + bx = 0$.
$$2t(t - 14) = 0$$ Therefore, factor.
$$2t = 0 \quad \text{or} \quad t - 14 = 0$$
$$t = 0 \quad \text{or} \quad t = 14$$

The solution set is $\{0, 14\}$.

12–5 *Methods of Solution* (continued)

> **d.** $n^2 + 6n - 16 = 0$ The equation has the form $x^2 + bx + c = 0$.
> $n^2 + 6n = 16$ Therefore, complete the square.
> $n^2 + 6n + 9 = 16 + 9$
> $(n + 3)^2 = 25$
> $n + 3 = \pm\sqrt{25}$
> $n = -3 \pm 5$
>
> The solution set is $\{-8, 2\}$.

Solve each quadratic equation using the most appropriate method.
Write irrational answers in simplest radical form. You may wish
to use a calculator.

1. $x^2 + 3x + 2 = 0$ **2.** $x^2 - 2x = 24$ **3.** $6x^2 = 96$

4. $6x^2 - 12x = 0$ **5.** $x^2 + 6x - 2 = 0$ **6.** $2x^2 + x = 10$

7. $x^2 + 5x = 6$ **8.** $2x^2 - 7x + 3 = 0$ **9.** $4x^2 - 36x = 0$

10. $(x - 2)^2 = 9$ **11.** $x^2 - 4x + 2 = 0$ **12.** $4x^2 + 4x - 3 = 0$

13. $x^2 - 4x = 12$ **14.** $x^2 - 11x + 24 = 0$ **15.** $x^2 + 3x = 18$

16. $x^2 - 2x = 1$ **17.** $m^2 - 2m = 35$ **18.** $2x^2 - 8x + 8 = 0$

19. $2x^2 + 5x = -2$ **20.** $3x^2 - 7x = 6$ **21.** $4x^2 + 4x - 15 = 0$

22. $6y^2 - 13y - 8 = 0$ **23.** $5n^2 + 14n - 3 = 0$ **24.** $2x^2 + 3x = 27$

25. $2x^2 - 9x + 7 = 0$ **26.** $\dfrac{1}{x} = \dfrac{2x - 5}{3}$ **27.** $\dfrac{x - 1}{2x + 1} = \dfrac{x + 1}{3x - 1}$

Mixed Review Exercises

Evaluate if $x = 1$, $y = 3$, and $z = -4$. Write irrational expressions in simplest
radical form.

1. $\pm\sqrt{y^2 - 4xz}$ **2.** $-\sqrt{y^2 + 4x}$ **3.** $\sqrt{z^2 - 4xy}$

4. $\sqrt{z^2 + 4xy}$ **5.** $\pm\sqrt{x^2 - 4yz}$ **6.** $\sqrt{x^2 + 4yz}$

Solve. Write irrational roots in simplest radical form.

7. $4x^2 + 4x - 15 = 0$ **8.** $2d^2 - 3d - 77 = 0$

9. $2(x + 1)(x + 3) = (2x + 1)(x + 5) - 2$ **10.** $c^2 - 3c - 1 = 0$

11. $3p^2 + 4p + 1 = 0$ **12.** $8y^2 + 14y = -3$

12–6 Solving Problems Involving Quadratic Equations

Objective: To use quadratic equations to solve problems.

Example   A landscaper wishes to design a rectangular formal garden that will be 6 m longer than the width. If the area of the garden is to be 135 m^2, find the length and the width.

Solution

Step 1   The problem asks for the length and the width of the garden.

Step 2   Let x = the width in meters.
Then $x + 6$ = the length in meters.

Step 3   Use the formula for the area of a rectangle to write an equation.

$$\text{Length} \times \text{Width} = \text{Area}$$
$$x(x + 6) = 135$$

Step 4   Solve.
$$x^2 + 6x = 135$$
$$x^2 + 6x + 9 = 135 + 9$$
$$(x + 3)^2 = 144$$
$$x + 3 = \pm \sqrt{144}$$
$$x + 3 = \pm 12$$
$$x = -3 \pm 12$$
$$x = 9 \quad \text{or} \quad x = -15$$

Step 5   Disregard the negative root since a negative length has no meaning.

Check: 9: $9(9 + 6) \stackrel{?}{=} 135$
$$9(15) \stackrel{?}{=} 135$$
$$135 = 135 \;\checkmark$$

The width of the garden is 9 m and the length is 15 m.

Solve. Give irrational roots to the nearest tenth. Use your calculator or a table of square roots as necessary.

1. The sum of a number and its square is 72. Find the number.

2. The sum of a number and its square is 30. Find the number.

3. The difference of a number and its square is 110. Find the number.

4. The difference of a number and its square is 132. Find the number.

5. The width of a rectangular garden is 4 m shorter than the length. If the area of the garden is 320 m^2, find the length and the width.

6. An architect wants to design a rectangular building such that the area of the floor is 400 yd^2. The length of the floor is to be 10 yd longer than the width. Find the length and the width of the floor.

12–6 *Solving Problems Involving Quadratic Equations* (continued)

Solve. Give irrational roots to the nearest tenth. Use your calculator or a table of square roots as necessary.

7. The length of a rectangle is 3 times the width. The area of the rectangle is 48 cm^2. Find the length and the width.

8. The length of a rectangular park is 5 m longer than the width. If the area of the park is 150 m^2, find the length and the width.

9. The length of a rectangle is twice the width. The area of the rectangle is 72 m^2. Find the length and the width.

10. The length of the base of a triangle is twice its altitude. If the area of the triangle is 144 cm^2, find the altitude. (*Hint:* Area of a triangle = $\frac{1}{2}$ × base × height.)

11. The altitude of a triangle is 5 m less than its base. The area of the triangle is 80 m^2. Find the base.

12. If the sides of a square are increased by 3 cm, its area becomes 144 cm^2. Find the length of the sides of the original square.

13. If the sides of a square are increased by 4 cm, its area becomes 196 cm^2. Find the length of the sides of the original square.

14. If you square a positive number, add ten times the number and subtract 84, the result is 180. What is the number?

15. Oscar has a rectangular garden that measures 18 m by 15 m. Next year he wants to increase the area to 340 m^2 by increasing the width and the length by the same amount. What will be the dimensions of the garden next year?

16. Working alone, Carrie can paint a house in 3 h less than Walter. Together Carrie and Walter can paint a house in 8 h. How long does it take Walter to paint a house alone?

17. Andrea can ride her bike 2 mi/h faster than Ruby. It takes Andrea 48 min less to travel 50 mi than it does Ruby. What is Andrea's rate in mi/h?

Mixed Review Exercises

Solve.

1. $\dfrac{\sqrt{n}}{2} = \dfrac{\sqrt{5}}{1}$

2. $\dfrac{\sqrt{x-2}}{4} = \dfrac{3}{2}$

3. $\dfrac{\sqrt{3t}}{8} = \dfrac{1}{4}$

4. $\dfrac{12}{5} = \dfrac{36}{n}$

5. $m^2 = 12m - 32$

6. $\dfrac{x-2}{x+3} + x = 2$

In Exercises 7–9, (x_1, y_1) and (x_2, y_2) are ordered pairs of the same direct variation. Find the missing value.

7. $x_1 = 1, y_1 = 3$
   $x_2 = 6, y_2 = \underline{\ ?\ }$

8. $x_1 = 8, y_1 = \underline{\ ?\ }$
   $x_2 = 4, y_2 = 5$

9. $x_1 = 12, y_1 = 16$
   $x_2 = \underline{\ ?\ }, y_2 = 8$

12–7 Direct and Inverse Variations Involving Squares

Objective: To use quadratic direct variation and inverse variation as a square in problem solving.

Vocabulary

Quadratic direct variation A function of the form

$$y = kx^2, \text{ where } k \text{ is a nonzero constant.}$$

You say that y *varies directly as x^2* or that y *is directly proportional to x^2*.
If (x_1, y_1) and (x_2, y_2) are ordered pairs of the same quadratic variation,
and neither x_1 nor x_2 is zero, then

$$\frac{y_1}{x_1^2} = \frac{y_2}{x_2^2}.$$

Inverse variation as the square A function of the form

$$x^2 y = k, \qquad \text{where } k \text{ is a nonzero constant,}$$
$$\text{or} \quad y = \frac{k}{x^2}, \qquad \text{where } x \neq 0.$$

You say that y *varies inversely as x^2* or y *is inversely proportional to x^2*.

If (x_1, y_1) and (x_2, y_2) are ordered pairs of the function defined by $x^2 y = k$, then

$$x_1^2 y_1 = x_2^2 y_2.$$

Example 1 Given that a varies directly as the square of d, and $a = 18$ when $d = 3$, find the value of d when $a = 8$.

Solution 1 Use $a = kd^2$: $18 = 9k$
$2 = k$
$a = 2d^2$

For $a = 8$: $8 = 2d^2$
$4 = d^2$
$\pm 2 = d$

Solution 2 $\dfrac{a_1}{d_1^2} = \dfrac{a_2}{d_2^2}$

$\dfrac{18}{3^2} = \dfrac{8}{d_2^2}$

$18 d_2^2 = 72$

$d_2^2 = 4$

$d_2 = \pm 2$

CAUTION Remember that in a particular situation you must check each root to see if it makes sense.

Solve. Give roots to the nearest tenth. You may wish to use a calculator.

1. Given that d varies directly as the square of t and $d = 16$ when $t = 2$, find the value of d when $t = 3$.

2. Given that A varies directly as the square of e and $A = 32$ when $e = 2$, find the value of e when $A = 8$.

3. The price of a diamond varies directly as the square of its mass in carats. If a 1.5 carat diamond costs $2700, find the cost of a 3 carat diamond.

12–7 Direct and Inverse Variations Involving Squares (continued)

Example 2   Given that h varies inversely as the square of r, and that $r = 3$ when $h = 18$, find the value of r when $h = \dfrac{3}{10}$.

Solution 1   Use $h = \dfrac{k}{r^2}$: $18 = \dfrac{k}{9}$

$162 = k$

$h = \dfrac{162}{r^2}$

For $h = \dfrac{3}{10}$: $\dfrac{3}{10} = \dfrac{162}{r^2}$

$3r^2 = 1620$

$r^2 = 540$

$r = \pm\sqrt{540}$

$ = \pm6\sqrt{15}$

Solution 2   Use $r_1^2 h_1 = r_2^2 h_2$

$(3)^2 18 = r_2^2\left(\dfrac{3}{10}\right)$

$162 = \dfrac{3r_2^2}{10}$

$3r_2^2 = 1620$

$r_2^2 = 540$

$r_2 = \pm\sqrt{540}$

$r_2 = \pm6\sqrt{15}$

Solve. Give roots to the nearest tenth. You may wish to use a calculator.

4. Given that h varies inversely as the square of r, and that $r = 4$ when $h = 25$, find the value of r when $h = 4$.

5. Given that h varies inversely as the square of r, and that $r = 5$ when $h = 36$, find the value of h when $r = 15$.

6. Light intensity varies inversely as the square of the distance from the source. At 2 m, an intensity scale has a reading of 6. What will be the reading at 6 m?

7. The distance it takes an automobile to stop varies directly as the square of its speed. If the stopping distance for a car traveling at 80 km/h is 175 m, what is the stopping distance for a car traveling at 48 km/h?

8. The height of a cylinder of a given volume is inversely proportional to the square of the radius. A cylinder of radius 4 cm has a height of 12 cm. What is the radius of a cylinder of equal volume whose height is 48 cm?

9. The length of a pendulum varies directly as the square of the time in seconds it takes to swing from one side to the other. If it takes a 100 cm pendulum 1 s to swing from one side to the other, how many seconds does it take a 50 cm pendulum to swing?

Mixed Review Exercises

Solve for the indicated variable. State any restrictions.

1. $S = C(1 - r)$; r

2. $d = \dfrac{a(2t - v)}{2}$; v

3. $A = \dfrac{h(b + c)}{2}$; c

Write an equation, in standard form, of the line passing through the given points.

4. (2, 5), (4, 6)

5. (0, 2), (−4, 1)

6. (2, 3), (−1, 0)

12–8 Joint and Combined Variation

Objective: To solve problems involving joint variation and combined variation.

Vocabulary

Joint variation When a variable varies directly as the product of two or more other variables. You can express the relationship in the forms

$$z = kxy, \quad \text{where } k \text{ is a nonzero constant,}$$

and

$$\frac{z_1}{x_1 y_1} = \frac{z_2}{x_2 y_2}.$$

You say that z *varies jointly as x and y.*

Combined variation When a variable varies directly as one variable and inversely as another. You can express the relationship in the forms

$$zy = kx \quad (\text{or } z = \frac{kx}{y}), \quad \text{where } k \text{ is a nonzero constant,}$$

and

$$\frac{z_1 y_1}{x_1} = \frac{z_2 y_2}{x_2}.$$

You say that z *varies directly as x and inversely as y.*

Example 1 The volume of a right circular cone varies jointly as the height, h, and the square of the radius, r. If $V_1 = 1848\pi$, $h_1 = 9$, $r_1 = 14$, $h_2 = 12$, and $r_2 = 7$, find V_2.

Solution

$$\frac{V_1}{h_1 r_1^2} = \frac{V_2}{h_2 r_2^2} \qquad \text{Use the form for joint variation.}$$

$$\frac{1848\pi}{9(14)^2} = \frac{V_2}{12(7)^2} \qquad \text{Substitute the given values.}$$

$$9(196)V_2 = 12(49)(1848\pi) \qquad \text{Solve for } V_2.$$

$$1764 V_2 = 1,086,624\pi$$

$$V_2 = 616\pi$$

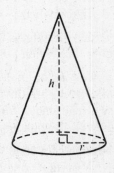

Solve. Give roots to the nearest tenth. You may wish to use a calculator.

1. d varies jointly as r and t. If $d = 75$ when $r = 30$ and $t = 10$, find d when $r = 18$ and $t = 6$.

2. m varies jointly as v and the square of u. If $m = 60$ when $v = 25$ and $u = 12$, find u when $m = 9$ and $v = 15$.

3. The volume of a pyramid varies jointly as the height and the area of the base. A pyramid 21 cm high has a base area of 25 cm^2 and a volume of 175 cm^3. What is the volume if the height is 15 cm and the area of the base is 36 cm^2?

2–8 Joint and Combined Variation (continued)

Example 2 The power, P, of an electric current varies directly as the square of the voltage, V, and inversely as the resistance, R. If 6 volts applied across a resistance of 3 ohms produces 12 watts of power, how much voltage applied across a resistance of 6 ohms will produce 24 watts of power?

Solution

$$\frac{P_1 R_1}{V_1^{\,2}} = \frac{P_2 R_2}{V_2^{\,2}}$$ Use the form for combined variation.

$$\frac{12(3)}{6^2} = \frac{24(6)}{V_2^{\,2}}$$ Let $P_1 = 12$, $V_1 = 6$, $R_1 = 3$, $P_2 = 24$, and $R_2 = 6$.

$$36V_2^{\,2} = 36(24)(6)$$ Solve for V_2.

$$V_2^{\,2} = 144$$ Discard the negative root since a negative

$$V_2 = \pm \sqrt{144} = \pm 12$$ voltage has no meaning.

12 volts will produce the required power.

olve. Give roots to the nearest tenth. You may wish to use a calculator.

4. c varies directly as a and inversely as b. If $c = 16$ when $a = 96$ and $b = 16$, find c when $a = 60$ and $b = 8$.

5. c varies inversely as the square of h and directly as n. If $c = 1$ when $h = 12$ and $n = 25$, find h when $c = 4.84$ and $n = 4$.

6. w varies directly as u and inversely as v^2. If $w = 8$ when $u = 2$ and $v = 3$, find u when $v = 2$ and $w = 27$.

7. If Q varies directly as the square of t and inversely as the cube of r, and $Q = 108$ when $t = 12$ and $r = 2$, find Q when $t = 15$ and $r = 3$.

8. The power, P, of an electric current varies directly as the square of the voltage, V, and inversely as the resistance, R. If 6 volts applied across a resistance of 3 ohms produces 12 watts of power, how much power will 12 volts applied across a resistance of 9 ohms produce?

9. The volume of a cone varies jointly as its height and the square of its radius. A certain cone has a volume of 154π cm^3, a height of 3 cm, and a radius of 7 cm. Find the radius of another cone that has a height of 7 cm and a volume of 264π cm^3.

0. The distance a car travels from rest varies jointly as its acceleration and the square of the time of motion. A car travels 500 m from rest in 5 s at an acceleration of 40 m/s^2. How many seconds will it take the car to travel 150 m at an acceleration of 3 m/s^2?

Mixed Review Exercises

olve each system.

1. $2x - y = -1$
$x - y = -2$

2. $x + y = 6$
$x - 2y = 12$

3. $2x - 5y = 2$
$-x + 5y = 4$

4. $3x + 2y = 3$
$x + 2y = 9$

Looking Ahead

Sample Spaces and Events

Objective: To specify the sample space and events for a random experiment.

Vocabulary

Probability The branch of mathematics that deals with the possibility, or
likelihood, that an event will happen.

Random experiment An activity repeated under essentially the same
conditions where the outcome of the activity can't be predicted.

Sample space of an experiment The set of all possible outcomes of a
random experiment.

Event Any subset of the sample set of an experiment.

Simple event An event involving a single member of the sample space.

Example 1   For the experiment of spinning the pointer of the spinner below, give:

   a. the sample space for the experiment.

   b. the event that an odd number results.

   c. the event that a number less than 4 results.

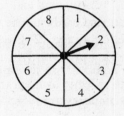

Solution   **a.** {1, 2, 3, 4, 5, 6, 7, 8}

   b. {1, 3, 5, 7}

   c. {1, 2, 3}

Example 2   Two spinners are divided as shown at the
right. For this experiment, give:

   a. the sample space for the experiment.

   b. the event that the sum of the numbers
   on the two spinners equals 5.

   c. the event that the sum of the numbers
   on the two spinners is more than 5.

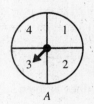 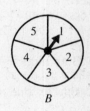

A   B

Solution   A simple event can be represented by an ordered pair (A, B), where A is the number
from spinner A and B is the number from spinner B.

   a. {(1, 1), (1, 2), (1, 3), (1, 4), (1, 5), (2, 1), (2, 2), (2, 3), (2, 4), (2, 5),
   (3, 1), (3, 2), (3, 3), (3, 4), (3, 5), (4, 1), (4, 2), (4, 3), (4, 4), (4, 5)}

   b. {(1, 4), (2, 3), (3, 2), (4, 1)}

   c. {(1, 5), (2, 4), (2, 5), (3, 3), (3, 4), (3, 5), (4, 2), (4, 3), (4, 4), (4, 5)}

Sample Spaces and Events   (continued)

For the experiments described in Exercises 1–5, first give the sample space and then give each event.

1. Each of the letters P, Q, R, S, T, U, V, W, and X is written on a card. The cards are shuffled, and then one card is drawn at random.

   a. The letter is a vowel.

   b. The letter is not a vowel.

   c. The letter is W or X.

2. A glass jar contains a yellow marble and a blue marble. A second jar contains a white, a green, a blue, and a red marble. One marble is taken at random from each jar.

   a. One marble is red.

   b. At least one marble is blue.

   c. Neither marble is green.

3. A hat contains cards numbered 1, 2, 4, and 8. A second hat contains cards numbered 3, 4, and 12. One card is drawn at random from each hat.

   a. Both numbers are the same.

   b. Both numbers are factors of 6.

   c. The sum of the numbers is less than 7.

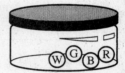

   d. The sum of the numbers is greater than 12.

4. A spinner is divided into four equal sections numbered 1, 2, 3, and 4. A second spinner is divided into three equal sections numbered 1, 7, and 14. The pointer on each spinner is spun.

   a. Both numbers are even.

   b. Exactly one number is even.

   c. The sum of the numbers is between 8 and 18.

   d. The product of the numbers is greater than 15.

5. Refer to the two-spinner experiment in Example 2. The ordered pair (a, b) represents a simple event.

   a. $a + b > 6$ **b.** $a < b$ **c.** $a + b$ is a multiple of 4.

   d. $a \cdot b = a + b$ **e.** $a + b = 7$ **f.** $a + b$ is an even number.

Probability

Objective: To find the probability that an event will occur.

Vocabulary

Probability of an event The ratio of the number of outcomes favoring the event to the total number of possible outcomes. For example, in the case of a tossed coin, the probability of $\{H\}$, written $P(H)$, is $\frac{1}{2}$ and the probability of $\{T\}$, written $P(T)$, is also $\frac{1}{2}$.

The probability of an impossible event is 0.

The probability of an event that is certain is 1.

For any probability P, $0 \leq P \leq 1$.

Example   A glass bowl contains 4 red marbles, 3 blue marbles, and 1 white marble. A marble is drawn at random from the bowl. Find the probability of each event.

    a. Event A: The marble drawn is red.

    b. Event B: The marble drawn is either red or blue.

    c. Event C: The marble drawn is not red.

    d. Event D: The marble drawn is green.

Solution   The sample space is {red 1, red 2, red 3, red 4, blue 1, blue 2, blue 3, white}.

    a. Since there are 4 red marbles, Event A has 4 equally likely outcomes. So $P(A) = \frac{4}{8} = \frac{1}{2}$.

    b. Since there are 4 red marbles and 3 blue marbles, Event B has 7 equally likely outcomes. So $P(B) = \frac{7}{8}$.

    c. If a marble is not red, then it must either be blue or white. Since there are 3 blue marbles and 1 white marble, Event C has 4 equally likely outcomes. So $P(C) = \frac{4}{8} = \frac{1}{2}$.

    d. Since there aren't any green marbles, $P(D)$ is an impossible event, so $P(D) = 0$.

Solve.

1. A jar contains 4 red marbles, 5 white marbles, and 3 blue marbles. A marble is drawn at random from the jar. Find the probability of each event.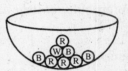

    a. The marble is red.   **b.** The marble is white.

    c. The marble is either red or white.   **d.** The marble is black.

    e. The marble is either red, white, or blue.

robability *(continued)*

lve.

. One card is drawn at random from a standard deck of 13 hearts, 13
diamonds, 13 clubs, and 13 spades. Find the probability of the event
that the card is:

a. an ace.   **b.** a black 2.

c. a spade.   **d.** the jack of diamonds.

e. a 2 or 3.   **f.** an ace, king, or queen.

. A letter is drawn at random from the word Q U A R T E R. Find the
probability of the event that the letter is:

a. a vowel.   **b.** a consonant.

c. a T.   **d.** an R.

. A cube whose sides are numbered 1, 2, 3, 4, 5, and 6 is rolled. Find
the probability of the event that the number on the cube is:

a. a 4.   **b.** a 3 or 6.

c. greater than 2.   **d.** greater than 1 but less than 4.

. A spinner has six equal sections numbered 1, 2, 3, 4, 5, and 6. The
pointer on the spinner is spun. Find the probability of each event.

a. The pointer stops on an odd number.

b. The pointer stops on a number less than 4.

c. The pointer stops on a multiple of 2.

d. The pointer stops on a perfect square.

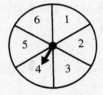

. A penny, a nickel, and a quarter are tossed. Find the probability of
the event that the coins show:

a. 3 tails.   **b.** exactly 2 tails.

c. at least 2 tails.   **d.** one or two tails.

. A spinner is divided into four equal sections numbered 1, 2, 3, and 4.
A second spinner is divided into three equal sections numbered 2, 4,
and 8. Each pointer is spun. Find the probability of each event.

a. $P(1, 8)$

b. $P(4, \text{not } 4)$

c. $P(4, 8)$

d. $P(\text{even number, even number})$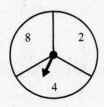

e. $P(\text{sum is } 6)$

f. $P(\text{sum is more than } 7)$

NAME _____ DATE _____

Frequency Distributions

Objective: To recognize and analyze frequency distribution.

Vocabulary

Frequency distribution A table summarizing data.

Histogram A graph that describes a frequency distribution.

Mean Sum of the data in a frequency distribution divided by the number of items in the data.

Median The middle number when the data in a frequency distribution are ranked in order.

Mode The most frequently occurring number in a frequency distribution. A frequency distribution may have more than one mode, or no mode.

Range The difference between the highest and lowest values in a frequency distribution.

Example 1 Make a frequency distribution and a histogram to show the price of a gallon of milk during the last fourteen weeks:

$$\$1.19, \$1.08, \$1.10, \$1.05, \$1.15, \$1.19, \$1.09,$$
$$\$1.10, \$1.15, \$1.15, \$1.10, \$1.15, \$1.15, \$1.08$$

Solution To make a frequency distribution, arrange the data in a table in increasing order and count how many times each price occurred as shown at the right.

To make a histogram, as shown below, group the data into convenient intervals. A "boundary" score is usually included in the interval to its left.

Price	Number of Weeks
$1.05	1
$1.08	2
$1.09	1
$1.10	3
$1.15	5
$1.19	2

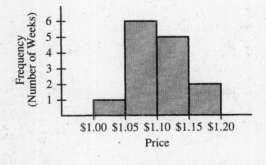

---

1. Make a histogram for the test scores received by twenty students:

94, 84, 95, 88, 74, 97, 83, 89, 90, 72, 78, 81, 67, 76, 84, 80, 75, 98, 92, 86

requency Distribution (continued)

Example 2 The top speeds in miles per hour of ten animals are listed in increasing order:

1, 18, 25, 32, 32, 35, 43, 50, 64, 70

Find the mean, the median, the mode, and the range of the data.

Solution $Mean = \dfrac{1 + 18 + 25 + 32 + 32 + 35 + 43 + 50 + 64 + 70}{10} = \dfrac{370}{10} = 37$

Since there is an even number of data, the median is the average of the two middle scores.

$Median = \dfrac{32 + 35}{2} = \dfrac{67}{2} = 33.5$

Since 32 is the score that occurs most frequently, it is the *mode*.

The *range* is the difference between 70 and 1, or 69.

---

or the data in Exercises 2–7, find (a) the mean, (b) the median,
the mode, and (d) the range. Whenever appropriate, give answers
the nearest tenth.

. 21, 50, 35, 49, 35   **3.** 8, 16, 43, 32, 16   **4.** 37, 36, 27, 38, 38, 32, 23

. 13, 8, 12, 8, 12, 8, 16   **6.** 51, 43, 42, 46, 51, 46, 53   **7.** 70, 76, 64, 46, 80, 48, 64

r Exercises 8–15 answer the question asked. Whenever
propriate, give answers to the nearest tenth.

. In six football games, the Tigers scored 28, 32, 21, 28, 42, and 12 points. Find the mean, the median, the mode, and the range of the team's scores.

. Find the mean and the range for the average monthly temperatures given in degrees Celsius: $-1°$, $-2°$, $6°$, $21°$, $26°$, $30.5°$, $32°$, $26.5°$, $20.5°$, $6°$, $3°$, $2.5°$.

. In ten basketball games, Lyn scores 10, 13, 12, 9, 11, 16, 17, 16, 12, and 20 points. To the nearest tenth of a point, find the mean, the median, the mode, and the range.

. In a class of 15 students, the test scores were 80, 97, 93, 98, 100, 90, 67, 78, 96, 92, 85, 87, 83, 83, and 91. Find the mean, the median, the mode, and the range.

. The mean of 5 numbers is 26. What is the sum of the numbers?

. Mona needs an average bowling score of 175 to bowl in a tournament. If she has scores of 160, 165, 172, and 168 in the first four games, what does she have to bowl in the fifth game to be eligible for the tournament?

. If each employee in a business gets a $15 per week increase in salary, how does this affect:
   a. mean salary?   **b.** mode?   **c.** median?   **d.** range?

. If each number in a set of data is doubled, how does this affect the:
   a. mean?   **b.** mode?   **c.** median?   **d.** range?

Presenting Statistical Data

Objective: To construct stem-and-leaf plots and box-and-whisker plots.

Vocabulary

First quartile score Median of the bottom half of data which are arranged in order.

Third quartile score Median of the top half of data which are arranged in order.

Example 1   Construct a **stem-and-leaf plot** for the following set of test scores.

75	79	88	67	89	56	94	92	82	83
92	89	94	97	86	68	58	74	48	76
68	95	91	50	93	82	64	97	95	46

Solution   First the stems, derived by dropping the unit's digit from each score, are written in order to the left of a vertical line. For each score the *leaf*, or unit's digit, is then recorded to the right of the corresponding *stem*. For the score of 94, for example, the leaf 4 is recorded to the right of the stem 9.

```
4 | 8, 6
5 | 6, 8, 0
6 | 7, 8, 8, 4
7 | 5, 9, 4, 6
8 | 8, 9, 2, 3, 9, 6, 2
9 | 4, 2, 2, 4, 7, 5, 1, 3, 7, 5
```

---

1. Use the distribution of the scores given by the stem-and-leaf plot shown at the right.

   a. List the scores of the distribution in order.

   b. Find the median, the mode, and the range of the scores.

   ```
   4 | 7, 5
   5 | 8, 3, 6, 4
   6 | 4, 1, 5, 8
   7 | 6, 2, 5, 8, 0, 2
   8 | 1, 6, 5
   9 | 7, 3
   ```

2. In an effort to raise money for a charity, twenty students were involved in a basketball free-throw competition. Each person shot 100 free-throws and recorded the number made:

23	41	65	64	95	26	53	50	76	88
30	36	48	85	84	39	70	90	62	76

   Construct a stem-and-leaf plot for the number of free-throws that were made.

3. The students in a marketing class reported the amount of money they earned last week in part-time jobs. They earned (in dollars):

113	126	98	102	91	84	125	154	148	124
118	110	82	104	132	125	143	92	77	120

   Construct a stem-and-leaf plot for their earnings.

---

resenting Statistical Data *(continued)*

Example 2   Construct a **box-and-whisker plot** for the data listed in Exercise 2.

Solution   List the numbers in order from smallest to largest. Then find the highest score, lowest score, median score, and first and third quartile scores.

```
        ┌──────── bottom half of data ────────┐        ┌──────────── top half of data ───────────┐
        23  26  30  36  39  41  48  50  53  62        64  65  70  76  76  84  85  88  90  95
        lowest           first                median                 third                highest
        score            quartile             score                  quartile             score
          ↓              score                  ↓                    score                  ↓
          23               40                   63                     80                   95
```

Identify these five special values with dots below a number line.

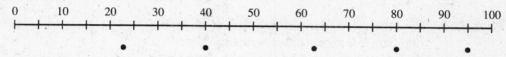

Next, make a *box* with the two quartile scores on the outer sides. Draw a line inside the box, through the median dot. Finally, draw *"whiskers"* from the sides of the box to the dots of the lowest and highest scores.

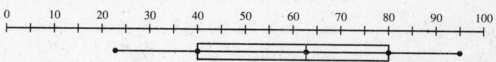

The box encloses the middle half of the data while the whiskers show the range.

The mean daily temperatures in degrees Fahrenheit for Springfield, as recorded on the fifteenth of each month, are listed:

    39°   44°   49°   56°   62°   69°   73°   74°   68°   60°   48°   44°

Find the:

a. median   **b.** first quartile temperature   **c.** third quartile temperature

d. highest temperature   **e.** lowest temperature   **f.** range

Construct a box-and-whisker plot for the data in Exercise 4.

Fifteen sales representatives listed the average amount they spent each night for lodging:

    $38 $74 $42 $56 $61 $48 $36 $40 $48 $60 $64 $52 $70 $50 $49

Find the:

a. median   **b.** first quartile amount   **c.** third quartile amount

d. highest amount   **e.** lowest amount   **f.** range

Make a box-and-whisker plot for the data in Exercise 6.

Points, Lines, and Angles

Objective: To represent points, lines, and angles and to measure angles.

Vocabulary

Point An exact location in space.

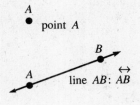
point A

Line A set of points that extends without end in opposite directions. Any two points determine a line.

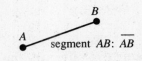
line AB: $\overleftrightarrow{AB}$

Line segment A part of a line that consists of two points and all the points between them. The length of $\overline{AB}$ is denoted by AB.

segment AB: $\overline{AB}$

Ray The part of a line that starts at a point A and extends without ending through a point B. A is called the *endpoint* of $\overrightarrow{AB}$.

ray AB: $\overrightarrow{AB}$

Angle A figure formed by two different rays that have the same endpoint. The rays are called the *sides* of the angle and the common endpoint is called the *vertex*. When three letters are used to name an angle, the middle letter is always the vertex.

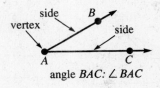
angle BAC: $\angle BAC$

Acute angle An angle which has measure between $0°$ and $90°$.

Right angle An angle which has measure $90°$.

Obtuse angle An angle which has measure between $90°$ and $180°$.

Straight angle An angle which has measure $180°$.

Protractor An instrument used to find the degree measure of an angle.

Example	Measure $\angle ABC$ using a protractor.
Solution	1. Put the center of the protractor on the vertex B.
	2. Place the $0°$ mark on side BC.
	3. Read the number where side BA crosses the scale.
	The measure of $\angle ABC$ is $40°$

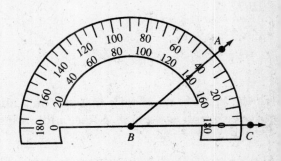

---

oints, Lines, and Angles (continued)

xercises 1–8 refer to the diagram at the right.

ate the measure of the angle.

1. ∠POT   2. ∠SOR

3. ∠QOU   4. ∠VOS

ame an angle whose measure is given.

5. 45°   6. 80°

7. 105°   8. 10°

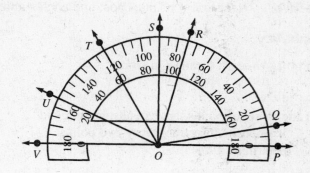

Exercises 9–12, name five different line segments in each diagram.

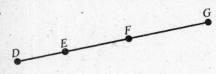

10.

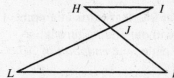

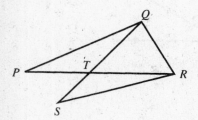

12.

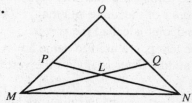

easure the given angle and classify it as acute, obtuse, right, or straight.

∠FOE

∠COF

∠AOH

∠AOG

∠FOG

∠BOF

∠FOA

∠AOC

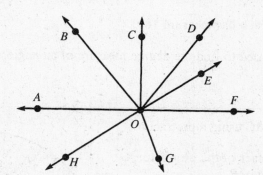

aph the solution set of each sentence on a number line.
ntify the graph as a point, a line, a line segment, or a ray.

$x = 1$ **22.** $x \geq 2$ **23.** $-3 \leq x \leq 0$

$2x - 3 = 1$ **25.** $-5 \leq x + 1 \leq 5$ **26.** $x < 1$ or $x > 0$

Pairs of Angles

Objective: To learn the names and properties of special pairs of angles.

Vocabulary

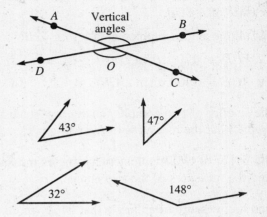

Vertical angles Two angles whose sides are rays in the same lines but in opposite directions. For example, in the diagram, $\angle AOB$ and $\angle COD$ are vertical angles as are $\angle AOD$ and $\angle COB$. Vertical angles have equal measures.

Complementary angles Two angles the sum of whose measures is 90°. Each angle is called a *complement* of the other.

Supplementary angles Two angles the sum of whose measures is 180°. Each angle is called a *supplement* of the other.

Example 1   The smaller of two complementary angles measures 30° less than the larger. Find the measure of the two angles.

Solution   Let n = the measure of the smaller angle in degrees.
Then $n + 30$ = the measure of the larger angle in degrees.

$$n + (n + 30) = 90$$
$$2n + 30 = 90$$
$$2n = 60$$
$$n = 30$$
$$n + 30 = 60$$

The check is left to you.

The measures of the angles are 30° and 60°.

Example 2   An angle measures 20° more than its supplement. Find the measure of the two angles.

Solution   Let x = the measure of the angle in degrees.
Then $180 - x$ = the measure of the angle's supplement in degrees.

$$x = 20 + (180 - x)$$
$$x = 20 + 180 - x$$
$$2x = 200$$
$$x = 100$$
$$180 - x = 80$$

The check is left to you.

The measure of the angle is 100°.
The measure of the supplement is 80°.

Pairs of Angles (continued)

In Exercises 1–4, use the diagram at the right below. Assume that the measures of ∠AGH and ∠DHG are equal.

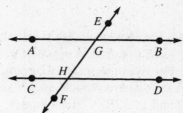

1. List all the angles with measures equal to the measure of ∠EGB.

2. List all the angles supplementary to ∠CHF.

3. If ∠FHD = 120°, then ∠FHC = _?_ .

4. If ∠AGE = 140°, then ∠CHG = _?_ .

5. The smaller of two complementary angles measures 20° less than the larger. Find the measures of the two angles.

6. The larger of two supplementary angles measures 5 times the smaller. Find the measures of the two angles.

7. Find the measure of an angle that is 62° more than the measure of its supplement.

8. Find the measure of an angle that is twice the measure of its complement.

9. Find the measure of an angle that is $\frac{1}{3}$ the measure of its supplement.

Example 3 The measure of a supplement of an angle is 30° more than four times the measure of its complement. Find the measure of the angle.

Solution Let n = the measure of the angle in degrees.
Then $90 - n$ = the measure of its complement, and
$180 - n$ = the measure of its supplement.

$$180 - n = 4(90 - n) + 30$$
$$180 - n = 360 - 4n + 30$$
$$3n = 210$$
$$n = 70$$

The measure of the complement is $(90 - 70)°$, or 20°.
The measure of the supplement is $(180 - 70)°$, or 110°.

Check: $110 \stackrel{?}{=} 4(20) + 30$
$110 = 110 \checkmark$

The measure of the angle is 70°.

---

10. The sum of the measures of a complement and a supplement of an angle is 154°. Find the measure of the angle.

11. The measure of a supplement of an angle is 20° more than twice the measure of its complement. Find the measure of the angle.

12. The measure of a supplement of an angle is four times the measure of its complement. Find the measure of the angle.

Triangles

Objective: To learn some properties of triangles.

Vocabulary

Triangle A figure formed by three segments joining three points not on the same line. Each segment is a *side* of the triangle. Each of the three points is a *vertex* of the triangle. *Angles* of the triangle are formed by two sides and a vertex. The sum of the angles of a triangle is 180°.

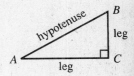

Right triangle A triangle having one right angle.

Isosceles triangle A triangle having two sides equal in length. The third side is called the *base*. The angles on either side of the base are called *base angles* and have equal measures.

Equilateral triangle A triangle with all sides of equal length. The angles of an equilateral triangle all measure 60°.

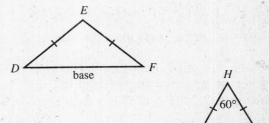

Example 1   The measures of two angles of a triangle are 26° and 64°. Find the measure of the third angle.

Solution   In $\triangle ABC$, $\angle A + \angle B + \angle C = 180°$.

Let x = the measure of the third angle.

$$x + 26 + 64 = 180$$
$$x + 90 = 180$$
$$x = 90$$

The check is left for you.

The third angle has a measure of 90°.

The measure of two angles of a triangle are given. Find the measure of the third angle.

1. 35°, 65°

2. 110°, 40°

3. 78°, 48°

4. 90°, 21°

5. 25°, 36°

6. 116°, 23°

Example 2   Use the converse of the Pythagorean theorem to determine whether a triangle with sides 8, 15, and 17 is a right triangle.

Solution $a^2 + b^2 = c^2$ (in a right triangle)

$$8^2 + 15^2 \overset{?}{=} 17^2$$
$$64 + 225 \overset{?}{=} 289$$
$$289 = 289 \ \checkmark \quad \text{A triangle with sides 8, 15, and 17 is a right triangle.}$$

Triangles (continued)

In Exercises 7–12, use the converse of the Pythagorean theorem to determine whether or not the triangle is a right triangle.

7. $\triangle ABC$: $AB = 16$, $BC = 12$, $AC = 20$

8. $\triangle DEF$: $EF = 29$, $FD = 21$, $DE = 20$

9. $\triangle GHI$: $GH = HI = 10$, $GI = 15$

10. $\triangle JKL$: $JK = 7$, $KL = 10$, $JL = 13$

11. $\triangle MNO$: $MN = 10$, $MO = 8$, $NO = 16$

12. $\triangle PQR$: $PQ = 16$, $QR = 34$, $PR = 30$

Example 3 If $\triangle EFG$ is a right triangle with $\angle F = 90°$, $EF = 15$, and $EG = 39$, find FG.

Solution Draw a sketch to help you solve the problem. Note that EG is the hypotenuse.

$$EF^2 + FG^2 = EG^2$$
$$15^2 + FG^2 = 39^2$$
$$225 + FG^2 = 1521$$
$$FG^2 = 1296$$
$$FG = 36$$

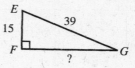

---

13. If $\triangle STU$ is a right triangle with $\angle T = 90°$, $ST = 12$, and $SU = 15$, find TU.

14. If $\triangle XYZ$ is a right triangle with $\angle Z = 90°$, $XZ = 24$, and $YZ = 10$, find XY.

15. If $\triangle ABC$ is a right triangle with $\angle A = 90°$, $AB = 9$, and $AC = 40$, find BC.

16. If $\triangle DEF$ is isosceles, $DE = DF$, and $\angle D = 46°$, find $\angle E$.

17. If $\triangle GHI$ is isosceles, $GH = GI$ and $\angle H = 30°$, find $\angle G$.

18. If $\triangle ABC$ is a right isosceles triangle and $\angle C = 90°$, find the measures of $\angle A$ and $\angle B$.

Exercises 19–24, $\angle C = 90°$ in $\triangle ABC$. Given the lengths of the other two sides, find the length of the third side in simplest radical form. Use the diagram shown below to help you.

19. $AC = 6$, $BC = 12$

20. $AC = 10$, $BC = 24$

21. $AC = 24$, $AB = 25$

22. $BC = 9$, $AC = 40$

23. $BC = 8$, $AB = 12$

24. $AC = 8$, $AB = 17$

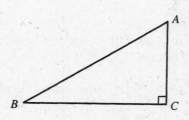

Similar Triangles

Objective: To solve problems involving similar triangles.

Vocabulary

Similar figures Two figures with the same shape are called similar.

Similar triangles Two triangles are similar when the measures of two angles of one triangle equal the measures of two angles of the other triangle. The triangles shown below are similar. Notice that the measures of the third set of angles are also equal.

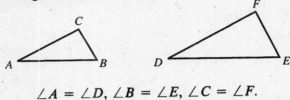

$$\angle A = \angle D, \ \angle B = \angle E, \ \angle C = \angle F.$$

Corresponding angles Angles with equal measures in similar triangles.

Corresponding sides The sides opposite corresponding angles in similar triangles. It is a geometric fact that the lengths of corresponding sides of similar triangles are proportional. For the triangles shown above:

$$\frac{AB}{DE} = \frac{BC}{EF} = \frac{CA}{FD}$$

In Exercise 1, refer to the diagram shown above in the Vocabulary section.

1. In $\triangle ABC$, $\angle A = 60°$ and $\angle B = 65°$.
   In $\triangle DEF$, $\angle E = 60°$ and $\angle F = 65°$.

   a. Write the corresponding angles.

   b. Write the corresponding sides.

   c. Complete: $\triangle ABC \sim \triangle \underline{\ ?\ }$.

2. In $\triangle JKL$ and $\triangle GHI$, $\angle J = \angle G = 50°$, and $\angle K = \angle H = 75°$.
   Write three equal ratios.

Example 1   In the diagram $\triangle ABC \sim \triangle DEF$.
Find AB and AC.

Solution   Corresponding sides are proportional:

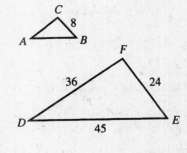

$$\frac{AB}{DE} = \frac{BC}{EF} \qquad \frac{AC}{DF} = \frac{BC}{EF}$$

$$\frac{AB}{45} = \frac{8}{24} \qquad \frac{AC}{36} = \frac{8}{24}$$

$$24(AB) = 360 \qquad 24(AC) = 288$$

$$AB = 15 \qquad\qquad AC = 12$$

Similar Triangles (continued)

Example 2   In the diagram, $\triangle ABC \sim \triangle AEF$.
Find EF and AF.

Solution   Corresponding sides are proportional:

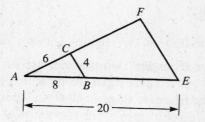

$$\frac{BC}{EF} = \frac{AB}{AE} \qquad \frac{AC}{AF} = \frac{AB}{AE}$$

$$\frac{4}{EF} = \frac{8}{20} \qquad \frac{6}{AF} = \frac{8}{20}$$

$$8(EF) = 80 \qquad 8(AF) = 120$$

$$EF = 10 \qquad AF = 15$$

Exercises 3–8, $\triangle ABC \sim \triangle DEF$. Find the lengths of the sides not given.

. $AB = 5$, $BC = 3$, $AC = 4$, $DE = 10$, $EF = \underline{\ ?\ }$, $DF = \underline{\ ?\ }$.

. $ED = 8$, $EF = 6$, $DF = 4$, $AC = 6$, $AB = \underline{\ ?\ }$, $BC = \underline{\ ?\ }$.

. $AB = 18$, $BC = 21$, $AC = 24$, $DE = 12$, $EF = \underline{\ ?\ }$, $FD = \underline{\ ?\ }$.

. $DE = 6$, $EF = 15$, $DF = 18$, $AC = 27$, $AB = \underline{\ ?\ }$, $BC = \underline{\ ?\ }$.

. $AB = 20$, $BC = 25$, $CA = 30$, $EF = 15$, $FD = \underline{\ ?\ }$, $DE = \underline{\ ?\ }$.

$ED = 7$, $EF = 6$, $FD = 4$, $CA = 12$, $BA = \underline{\ ?\ }$, $BC = \underline{\ ?\ }$.

the diagram, $\triangle ABC \sim \triangle AEF$.   **In the diagram, $\triangle ABC \sim \triangle ADE$.**

Find EF.   **11.** Find DE.

Find AF.   **12.** Find AC.

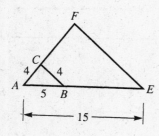

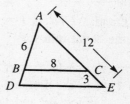

he diagram, $\triangle ABC \sim \triangle EBD$.

Find BE

Find BD.

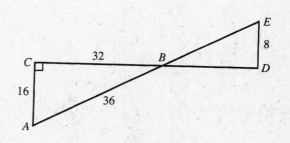

Trigonometric Ratios

Objective: To find the sine, cosine, and tangent of an acute angle.

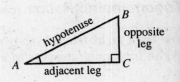

Vocabulary

Trigonometric ratios Ratios of the lengths of the sides of a right triangle.

The trigonometric ratios have special names. In right triangle ABC,   **Symbols**

sine of $\angle A$	$=$	$\dfrac{\text{length of leg opposite } \angle A}{\text{length of hypotenuse}}$	$= \dfrac{BC}{AB}$	$\sin A$
cosine of $\angle A$	$=$	$\dfrac{\text{length of leg adjacent to } \angle A}{\text{length of hypotenuse}}$	$= \dfrac{AC}{AB}$	$\cos A$
tangent of $\angle A$	$=$	$\dfrac{\text{length of leg opposite } \angle A}{\text{length of leg adjacent to } \angle A}$	$= \dfrac{BC}{AC}$	$\tan A$

Trigonometric functions The trigonometric ratios thought of as the values of three functions ($\sin A$, $\cos A$, $\tan A$) each having the set of acute angles as its domain.

Example 1   Find the sine, cosine, and tangent of $\angle A$ and $\angle B$.

Solution $\sin A = \dfrac{5}{13}$ $\sin B = \dfrac{12}{13}$

$\cos A = \dfrac{12}{13}$ $\cos B = \dfrac{5}{13}$

$\tan A = \dfrac{5}{12}$ $\tan B = \dfrac{12}{5}$

Example 2   Find the sine, cosine, and tangent of $\angle Q$.

Solution   Use the Pythagorean theorem to find RQ.

$a^2 + 10^2 = 18^2$

$a^2 + 100 = 324$

$a^2 = 224$

$a = 4\sqrt{14}$

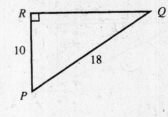

$\sin Q = \dfrac{10}{18} = \dfrac{5}{9}$ $\cos Q = \dfrac{4\sqrt{14}}{18} = \dfrac{2\sqrt{14}}{9}$ $\tan Q = \dfrac{10}{4\sqrt{14}} = \dfrac{5\sqrt{14}}{28}$

CAUTION   The trigonometric ratios are defined only for acute angles in a right triangle.

Trigonometric Ratios (continued)

For each right triangle shown, find sin A, cos A, tan A, sin B, cos B, and tan B.
Write irrational answers in simplest radical form.

1.

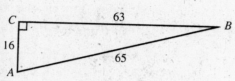

2.

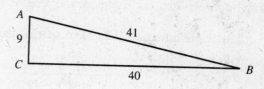

3.

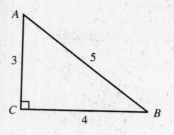

4.

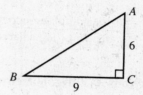

6.

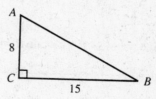

8.

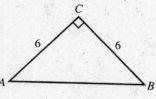

10.

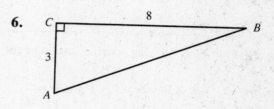

12.

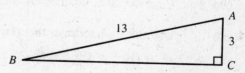

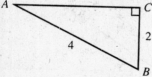

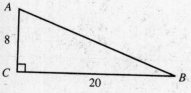

Values of Trigonometric Ratios

Objective: To find values of trigonometric ratios for given angles, and measures of angles for given trigonometric ratios.

Example 1   Find the values of:

      a. sin 56°   **b.** cos 56°   **c.** tan 56°.

Solution 1   **Using a Table**

It is not possible to give *exact* trigonometric ratios for most angles. You can use a table or a scientific calculator to find approximate values of trigonometric ratios.

Locate 56° in the left-hand column of the portion of the table shown at the right, then read across the row to find:

a. sin 56° ≈ 0.8290

b. cos 56° ≈ 0.5592

c. tan 56° ≈ 1.4826

For convenience, you may write = instead of ≈.

Angle	Sine	Cosine	Tangent
1°	.0175	.9998	.0175
56°	.8290	.5592	1.4826
57°	.8387	.5446	1.5399
58°	.8480	.5299	1.6003
59°	.8572	.5150	1.6643
60°	.8660	.5000	1.7321

Solution 2   **Using a Calculator**

a. To find the value for sin 56°, first enter 56. Then press the sin key to get 0.8290376.

    To the nearest ten-thousandth, sin 56° = 0.8290.

b. Enter 56 and then press the cos key to get:

    cos 56° ≈ 0.5591929, or cos 56° = 0.5592.

c. Enter 56 and then press the tan key to get:

    tan 56° ≈ 1.482561, or tan 56° = 1.4826.

Use a calculator or a table to find sin A, cos A, and tan A for the given measure of angle A.

1. 12°   **2.** 42°   **3.** 75°   **4.** 24°   **5.** 35°

6. 66°   **7.** 16°   **8.** 79°   **9.** 31°   **10.** 52°

11. 8°   **12.** 58°   **13.** 70°   **14.** 48°   **15.** 30°

---

alues of Trigonometric Ratios (continued)

Example 2   Find the measure of $\angle A$ to the nearest degree.

      a. $\cos A = 0.8290$ **b.** $\sin A = 0.7320$.

Solution 1   **Using a Table**

    a. Locate the value 0.8290 in the cosine column, if possible.

    Since it is there read across the row to the left-hand column to find that the angle has a measure of 34°.

    b. Since 0.7320 is not listed in the sine column, locate the entries between which 0.7320 lies.

    Since $\sin 47° = 0.7314$ and $\sin 48° = 0.7431$, the measure of $\angle A$ must be between 47° and 48°.

    Since 0.7320 is closer to 0.7314 than it is to 0.7431, $\angle A = 47°$ to the nearest degree.

Angle	Sine	Cosine	Tangent
1°	.0175	.9998	.0175
31°	.5150	.8572	.6009
32°	.5299	.8480	.6249
33°	.5446	.8387	.6494
34°	.5592	.8290	.6745
35°	.5736	.8192	.7002
46°	.7193	.6947	1.0355
47°	.7314	.6820	1.0724
48°	.7431	.6691	1.1106
49°	.7547	.6561	1.1504
50°	.7660	.6428	1.1918

Solution 2   **Using a Calculator**

Most calculators have the inverse keys ($\sin^{-1}$, $\cos^{-1}$, $\tan^{-1}$, or inv sin, inv cos, inv tan) that give the measure of an acute angle.

    a. Enter 0.8290, then press the $\cos^{-1}$ key to get 34.00385. To the nearest degree, $\angle A = 34°$.

    b. Enter 0.7320, then press the $\sin^{-1}$ key to get 47.054324. To the nearest degree, $\angle A = 47°$.

Use a calculator or a table to find the measure of angle A to the nearest degree.

6. $\sin A = 0.9903$

17. $\cos A = 0.9063$

8. $\cos A = 0.6428$

19. $\tan A = 1.7321$

20. $\tan A = 0.4456$

21. $\sin A = 0.6558$

22. $\tan A = 1.2851$

23. $\cos A = 0.9607$

24. $\tan A = 3.2608$

25. $\sin A = 0.9560$

26. $\sin A = 0.7777$

27. $\cos A = 0.6683$

28. $\sin A = 0.8330$

29. $\sin A = 0.6440$

30. $\tan A = 5.6708$

31. $\cos A = 0.0860$

Problem Solving Using Trigonometry

Objective: To use trigonometric ratios to solve problems.

Trigonometric ratios can be used to solve practical problems involving right triangles. You can find values for these ratios by using a table or a scientific calculator.

Example 1 An observation tower is 75 m high. A support wire is attached to the tower 20 m from the top. If the support wire and the ground form an angle of 46°, what is the length of the support wire?

Solution Draw a triangle and label the different values. First, find how high on the tower the support wire is attached.

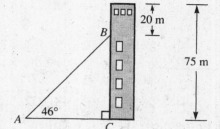

$$75 - 20 = 55$$

Since $\triangle ABC$ is a right triangle,

$$\sin 46° = \frac{55}{x}, \text{ or } x = \frac{55}{\sin 46°}.$$

From a table or a calculator, $\sin 46° = 0.7193$.

Then $x = \frac{55}{0.7193} \approx 76.5.$

To the nearest tenth of a meter, the support wire is 76.5 m long.

Example 2 A 12 ft ladder rests against the side of a building. If the foot of the ladder is 4 ft from the building, find, to the nearest degree, the angle the ladder makes with the ground.

Solution Draw a triangle and label the different values. Since $\triangle DEF$ is a right triangle,

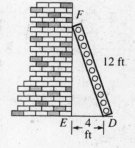

$$\cos \angle D = \frac{DE}{DF}$$

$$\cos \angle D = \frac{4}{12}$$

$$\cos \angle D = 0.3333$$

Therefore $\angle D = 71°.$

Use a table or calculator as needed to find the value of x to the nearest whole number.

1.

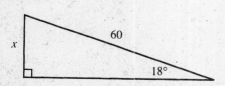

2.

roblem Solving Using Trigonometry *(continued)*

se a table or calculator as needed to find the value of *x* to the nearest
hole number.

3.

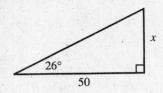

4.

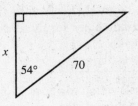

5.

6.

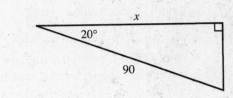

n a right △ *ABC*, ∠ *C* = 90°. Find the lengths of the other sides of
he triangle to the nearest whole number.

7. ∠*A* = 35°, *AB* = 48

8. ∠*B* = 23°, *AB* = 16

9. ∠*A* = 56°, *BC* = 40

10. ∠*A* = 44°, *BC* = 30

1. ∠*A* = 11°, *AC* = 17

12. ∠*B* = 30°, *AB* = 62

olve each problem, drawing a sketch for each. Express distances to
he nearest unit and angle measures to the nearest degree. Use a table
r a calculator as needed.

13. How far is it across the river?

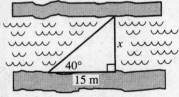

14. How long is the cable that supports the pole?

15. How high is the cliff?

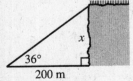

16. How high is the water ski jump?

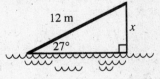

17. A kite is flying 115 ft above the ground. The length of the string to
the kite is 150 ft, measured from the ground. Find the angle to the
nearest degree that the string makes with the ground.

18. To the nearest degree what is the angle formed with the ground by a
32 ft ladder if it is leaning against a wall at a height of 28 ft?

Glossary of Properties

addition property of equality (p. 37): If a, b, and c are any real numbers, and $a = b$, then $a + c = b + c$ and $c + a = c + b$.

addition property of order (p. 169): For all real numbers a, b, and c:
1. If $a < b$, then $a + c < b + c$;
2. If $a > b$, then $a + c > b + c$.

associative properties (p. 19): For all real numbers a, b, and c:
$$(a + b) + c = a + (b + c)$$
$$(ab)c = a(bc)$$

closure properties (p. 19): For all real numbers a and b:

$a + b$ is a unique real number.
ab is a unique real number.

commutative properties (p. 19): For all real numbers a and b:
$$a + b = b + a$$
$$ab = ba$$

density property for rational numbers (p. 183): Between every pair of different rational numbers there is another rational number.

distributive property (of multiplication with respect to addition) (p. 27): For all real numbers a, b, and c,
$$a(b + c) = ab + ac \text{ and}$$
$$(b + c)a = ba + ca.$$

distributive property (of multiplication with respect to subtraction) (p. 27): For all real numbers a, b, and c,
$$a(b - c) = ab - ac \text{ and}$$
$$(b - c)a = ba - ca.$$

division property of equality (p. 39): If a and b are any real numbers, c is any nonzero real number, and $a = b$, then $\frac{a}{c} = \frac{b}{c}$.

identity property of addition (p. 21): There is a unique real number 0 such that for every real number a, $a + 0 = a$ and $0 + a = a$.

identity property of multiplication (p.29): There is a unique real number 1 such that for every real number a, $a \cdot 1 = a$ and $1 \cdot a = a$.

multiplication property of equality (p. 39): If a, b, and c are any real numbers, and $a = b$, then $ca = cb$ and $ac = bc$.

multiplication property of order (p.169): For all real numbers a, b, and c such that $c > 0$:
1. If $a < b$, then $ac < bc$;
2. If $a > b$, then $ac > bc$.
For all real numbers a, b, and c such that $c < 0$:
1. If $a < b$, then $ac > bc$;
2. If $a > b$, then $ac < bc$.

multiplicative property of -1 (p. 29): For every real number a, $a(-1) = -a$ and $(-1)a = -a$.

multiplicative property of zero (p. 29): For every real number a, $a \cdot 0 = 0$ and $0 \cdot a = 0$.

order property of rational number (p. 183): For all integers a and b and all positive integers c and d:
1. $\frac{a}{c} > \frac{b}{d}$ if and only if $ad > bc$.
2. $\frac{a}{c} < \frac{b}{d}$ if and only if $ad < bc$.

product property of square roots (p. 187): For any nonnegative real numbers a and b, $\sqrt{ab} = \sqrt{a} \cdot \sqrt{b}$.

property of comparison (p. 169): For all real numbers a and b, one and only one of the following statements is true:
$a < b, a = b, a > b$.

property of completeness (p. 189): Every decimal represents a real number, and every real number can be represented by a decimal.

property of the opposite of a sum (p. 21): For all real numbers a and b:
$$-(a + b) = (-a) + (-b).$$

property of opposites (p.21): For every real number a, there is a unique real number $-a$ such that $a + (-a) = 0$ and $(-a) + a = 0$.

property of opposites in products (p. 29): For all real numbers a and b,
$$(-a)(b) = -ab, a(-b) = -ab, \text{ and}$$
$$(-a)(-b) = ab.$$

property of quotients (p.75): If a, b, c, and d are real numbers with $b \neq 0$ and $d \neq 0$, then $\frac{ac}{bd} = \frac{a}{b} \cdot \frac{c}{d}$.

property of the reciprocal of the opposite of a number (p. 33): For every nonzero number a, $\frac{1}{-a} = -\frac{1}{a}$.

property of the reciprocal of a product (p. 33): For all nonzero numbers a and b,
$$\frac{1}{ab} = \frac{1}{a} \cdot \frac{1}{b}.$$

property of reciprocals (p. 33): For every nonzero real number a, there is a unique real number $\frac{1}{a}$ such that $a \cdot \frac{1}{a} = 1$ and $\frac{1}{a} \cdot a = 1$.

property of square roots of equal numbers (p. 191): For any real numbers r and s: $r^2 = s^2$ if and only if $r = s$ or $r = -s$.

quotient property of square roots (p. 187): For any nonnegative real number a and any positive real number b, $\sqrt{\frac{a}{b}} = \frac{\sqrt{a}}{\sqrt{b}}$.

reflexive property of equality: If a is a real number, then $a = a$.

substitution principle (p. 51): An expression may be replaced by another expression that has the same value.

subtraction property of equality (p.37): If a, b, and c are any real numbers, and $a = b$, then $a - c = b - c$.

symmetric property of equality: If a and b are real numbers, and $a = b$, then $b = a$.

transitive property of equality: For all real numbers, a, b, and c, if $a = b$ and $b = c$, then $a = c$.

transitive property of order (p. 169): For all real numbers a, b, and c,
1. If $a < b$ and $b < c$, then $a < c$;
2. If $c > b$ and $b > a$, then $c > a$.

zero-product property (p. 95): For all real numbers a and b, $ab = 0$ if and only if $a = 0$ or $b = 0$.